SHARING THE ADVENTURE WITH THE STUDENT

Exploring the Intersections of NASA Space Science and Education

A WORKSHOP SUMMARY

Dwayne Day, *Rapporteur*

Space Studies Board

Division on Engineering and Physical Sciences

Board on Science Education

NATIONAL RESEARCH COUNCIL
OF THE NATIONAL ACADEMIES

THE NATIONAL ACADEMIES PRESS
Washington, D.C.
www.nap.edu

THE NATIONAL ACADEMIES PRESS **500 Fifth Street, NW** **Washington, DC 20001**

NOTICE: The project that is the subject of this report was approved by the Governing Board of the National Research Council, whose members are drawn from the councils of the National Academy of Sciences, the National Academy of Engineering, and the Institute of Medicine.

This is a report of work supported by Contract NNH11CD57B between the National Academy of Sciences and the National Aeronautics and Space Administration. Any opinions, findings, or conclusions expressed in this publication are those of the author(s) and do not necessarily reflect the view of the organizations or agencies that provided support for the project.

International Standard Book Number-13: 978-0-309-37426-2
International Standard Book Number-10: 0-309-37426-X

Cover: Design by Tim Warchocki.

Limited copies of this report are available from:

Space Studies Board
National Research Council
500 Fifth Street, NW
Washington, DC 20001

Additional copies of this report are available from the National Academies Press, 500 Fifth Street, NW, Keck 360, Washington, DC 20001; (800) 624-6242 or (202) 334-3313; http://www.nap.edu.

Printed in the United States of America

THE NATIONAL ACADEMIES

Advisers to the Nation on Science, Engineering, and Medicine

The **National Academy of Sciences** is a private, nonprofit, self-perpetuating society of distinguished scholars engaged in scientific and engineering research, dedicated to the furtherance of science and technology and to their use for the general welfare. Upon the authority of the charter granted to it by the Congress in 1863, the Academy has a mandate that requires it to advise the federal government on scientific and technical matters. Dr. Ralph J. Cicerone is president of the National Academy of Sciences.

The **National Academy of Engineering** was established in 1964, under the charter of the National Academy of Sciences, as a parallel organization of outstanding engineers. It is autonomous in its administration and in the selection of its members, sharing with the National Academy of Sciences the responsibility for advising the federal government. The National Academy of Engineering also sponsors engineering programs aimed at meeting national needs, encourages education and research, and recognizes the superior achievements of engineers. Dr. C. D. Mote, Jr., is president of the National Academy of Engineering.

The **Institute of Medicine** was established in 1970 by the National Academy of Sciences to secure the services of eminent members of appropriate professions in the examination of policy matters pertaining to the health of the public. The Institute acts under the responsibility given to the National Academy of Sciences by its congressional charter to be an adviser to the federal government and, upon its own initiative, to identify issues of medical care, research, and education. Dr. Victor J. Dzau is president of the Institute of Medicine.

The **National Research Council** was organized by the National Academy of Sciences in 1916 to associate the broad community of science and technology with the Academy's purposes of furthering knowledge and advising the federal government. Functioning in accordance with general policies determined by the Academy, the Council has become the principal operating agency of both the National Academy of Sciences and the National Academy of Engineering in providing services to the government, the public, and the scientific and engineering communities. The Council is administered jointly by both Academies and the Institute of Medicine. Dr. Ralph J. Cicerone and Dr. C. D. Mote, Jr., are chair and vice chair, respectively, of the National Research Council.

www.national-academies.org

OTHER RECENT REPORTS OF THE SPACE STUDIES BOARD

Optimizing the U.S. Ground-Based Optical and Infrared Astronomy System (Board on Physics and Astronomy [BPA] and Space Studies Board [SSB], 2015)
Evaluation of the Implementation of WFIRST/AFTA in the Context of New Worlds, New Horizons in Astronomy and Astrophysics (BPA and SSB, 2014)
Pathways to Exploration: Rationales and Approaches for a U.S. Program of Human Space Exploration (Aeronautics and Space Engineering Board [ASEB], SSB, and the Committee on National Statistics, 2014)
Landsat and Beyond: Sustaining and Enhancing the Nation's Land Imaging Program (SSB, 2013)
Lessons Learned in Decadal Planning in Space Science: Summary of a Workshop (SSB, 2013)
Review of the Draft 2014 Science Mission Directorate Science Plan (SSB, 2013)
Solar and Space Physics: A Science for a Technological Society (SSB, 2013)
Assessment of a Plan for U.S. Participation in Euclid (BPA with SSB, 2012)
Assessment of Planetary Protection Requirements for Spacecraft Missions to Icy Solar System Bodies (SSB, 2012)
Earth Science and Applications from Space: A Midterm Assessment of NASA's Implementation of the Decadal Survey (SSB, 2012)
The Effects of Solar Variability on Earth's Climate: A Workshop Report (SSB, 2012)
NASA's Strategic Direction and the Need for a National Consensus (Division on Engineering and Physical Sciences, 2012)
Report of the Panel on Implementing Recommendations from the New Worlds, New Horizons Decadal Survey (BPA and SSB, 2012)
Technical Evaluation of the NASA Model for Cancer Risk to Astronauts Due to Space Radiation (SSB, 2012)
Assessment of Impediments to Interagency Collaboration on Space and Earth Science Missions (SSB, 2011)
Panel Reports—New Worlds, New Horizons in Astronomy and Astrophysics (BPA and SSB, 2011)
Recapturing a Future for Space Exploration: Life and Physical Sciences Research for a New Era (SSB and ASEB, 2011)
Sharing the Adventure with the Public—The Value and Excitement of "Grand Questions" of Space Science and Exploration: Summary of a Workshop (SSB, 2011)
Vision and Voyages for Planetary Science in the Decade 2013-2022 (SSB, 2011)
Capabilities for the Future: An Assessment of NASA Laboratories for Basic Research (Laboratory Assessments Board with SSB and ASEB, 2010)
Controlling Cost Growth of NASA Earth and Space Science Missions (SSB, 2010)
Defending Planet Earth: Near-Earth-Object Surveys and Hazard Mitigation Strategies: Final Report (SSB with ASEB, 2010)
An Enabling Foundation for NASA's Space and Earth Science Missions (SSB, 2010)
Forging the Future of Space Science: The Next 50 Years (SSB, 2010)
New Worlds, New Horizons in Astronomy and Astrophysics (BPA and SSB, 2010)
Revitalizing NASA's Suborbital Program: Advancing Science, Driving Innovation, and Developing a Workforce (SSB, 2010)

OTHER RECENT REPORTS OF THE BOARD ON SCIENCE EDUCATION

Reaching Students: What Research Says About Effective Instruction in Undergraduate Science and Engineering (Board on Science Education [BOSE], 2015)
Guide to Implementing the Next Generation Science Standards (BOSE, 2015)
Literacy for Science: Exploring the Intersection of the Next Generation Science Standards and the Common Core for ELA Standards: A Workshop Summary (BOSE, 2014)
Climate Change Education: Engaging Family Private Forest Owners on Issues Related to Climate Change: A Workshop Summary (BOSE, 2014)
Climate Change Education: Preparing Future and Current Business Leaders: A Workshop Summary (2014)
Developing Assessments for the Next Generation Science Standards (BOSE with Board on Testing and Assessment [BOTA], 2014)
STEM Integration in K-12 Education: Status, Prospects, and an Agenda for Research (BOSE and the National Academy of Engineering, 2014)
Monitoring Progress Toward Successful K-12 STEM Education: A Nation Advancing? (BOSE and BOTA, 2013)
Next Generation Science Standards: For States, By States (BOSE, 2013)

PLANNING COMMITTEE ON SHARING THE ADVENTURE WITH THE STUDENT: EXPLORING THE INTERSECTIONS OF NASA SPACE SCIENCE AND EDUCATION: A WORKSHOP

PHILIP R. CHRISTENSEN, Arizona State University, *Co-Chair*
BRETT D. MOULDING, Utah Partnership for Effective Science Teaching and Learning, *Co-Chair*
ALBERT BYERS, National Science Teachers Association
HEIDI B. HAMMEL, Association of Universities for Research in Astronomy
WESLEY L. HARRIS, Massachusetts Institute of Technology
CHARLES F. KENNEL, University of California, San Diego
JAMES MANNING, Consultant
RICHARD A. McCRAY, University of California, Berkeley
MITCHELL NATHAN, University of Wisconsin, Madison
PATRICIA H. REIFF, Rice University
THERESA SCHWERIN, Institute for Global Environmental Strategies

Staff

ABIGAIL SHEFFER, Program Officer
KATIE DAUD, Research Associate
ANESIA WILKS, Senior Program Assistant
MICHELLE THOMPSON, Lloyd V. Berkner Space Policy Intern
ANGELA DAPREMONT, Lloyd V. Berkner Space Policy Intern

MICHAEL H. MOLONEY, Director, Space Studies Board

SPACE STUDIES BOARD

BOARD ON SCIENCE EDUCATION

Acknowledgment of Reviewers

This workshop summary has been reviewed in draft form by individuals chosen for their diverse perspectives and technical expertise, in accordance with procedures approved by the NRC's Report Review Committee. The purpose of this independent review is to provide candid and critical comments that will assist the institution in making its published summary as sound as possible and to ensure that the summary meets institutional standards for clarity, objectivity, and responsiveness to the charge. The review comments and draft manuscript remain confidential to protect the integrity of the process. We wish to thank the following individuals for their review of this workshop summary:

Bonnie Eisenhamer, Space Telescope Science Institute,
Neil Gehrels, NASA Goddard Space Flight Center,
Michael Lach, University of Chicago, and
Mordecai Mac Low, American Museum of Natural History.

Although the reviewers listed above have provided many constructive comments and suggestions, they did not see the final draft of the workshop summary before its release. The review of this summary was overseen by James S. Trefil, George Mason University. Appointed by the National Research Council, he was responsible for making certain that an independent examination of this summary was carried out in accordance with institutional procedures and that all review comments were carefully considered. Responsibility for the final content of this summary rests entirely with the author(s) and the institution.

Contents

Introduction and Background

On December 2-3, 2014, the Space Studies Board and the Board on Science Education of the National Research Council (NRC) held a workshop on the NASA Science Mission Directorate (SMD) education program—"Sharing the Adventure with the Student." The discussion of NASA SMD's education efforts is particularly timely because of recent changes in K-12 science education policy and practices and a proposed reorganization of all of NASA SMD's science, technology, engineering, and mathematics (STEM) education efforts.

"Sharing the Adventure with the Student: Exploring the Intersections of NASA Space Science and Education—A Workshop" was organized by an ad hoc committee under the auspices of members from the Space Studies Board, serving as representatives of the space science community; the Board on Science Education, serving as representatives of experts in the creation and evaluation of STEM education efforts; as well as other experts. The workshop brought together these respective communities to promote a new dialog with the aim of increasing mutual understanding of how best to translate space science into useful educational materials and experiences.

This workshop summary has been prepared by the workshop rapporteur as a factual summary of what occurred at the workshop. The planning committee's role was limited to planning and convening the workshop. The views contained in the report are those of individual workshop participants and do not necessarily represent the views of the workshop participants as a whole, the planning committee, or the NRC.

This is the second in a series of workshops on NASA science communication and education. Previously, on November 8-10, 2010, the Space Studies Board held a public workshop, "Sharing the Adventure with the Public,"[1] that brought together scientists and professional communicators to discuss how NASA and its associated science and exploration communities can be more effective in communicating with the public.[2] The 2010 workshop participants discussed examples of where communication with the public has been challenging—such as for climate change—and where communication can be used more effectively to increase public support for space science. Science journalists offered tips for improving scientists' communication—such as becoming more active on social media sites. The gathering together of these communities in itself helped to improve communication in science, with all groups leaving the workshop with a better understanding of each other.

[1] National Research Council, *Sharing the Adventure with the Public: The Value and Excitement of "Grand Questions" of Space Science and Exploration: Summary of a Workshop*, The National Academies Press, Washington, D.C., 2011.

[2] More information and video recordings of Sharing the Adventure with the Public: The Value and Excitement of "Grand Questions" of Space Science and Exploration are available at http://sites.nationalacademies.org/SSB/CompletedProjects/SSB_065881.

THE BACKGROUND OF NASA EDUCATION EFFORTS

The National Aeronautics and Space Act of 1958, which created NASA, directed that the agency should pursue several goals. Among these are the following:

- The expansion of human knowledge of Earth and of phenomena in the atmosphere and space; and
- The preservation of the role of the United States as a leader in aeronautical and space science and technology and in the application thereof to the conduct of peaceful activities within and outside the atmosphere.

NASA has interpreted these goals to include support for the goals of American educational institutions at all levels. A 2008 NRC report, *NASA's Elementary and Secondary Education Program: Review and Critique,*[3] recommended the following:

> NASA should continue to engage in education activities at the K-12 level, designing its K-12 education activities so that they capitalize on NASA's primary strengths and resources, which are found in the mission directorates. These strengths and resources are the agency's scientific discoveries; its technology and aeronautical developments; its space exploration activities; the scientists, engineers, and other technical staff (both internal and external) who carry out NASA's work; and the unique excitement generated by space flight and space exploration (p. 6).

The report also noted that among the large number of agency staff who focus on science, engineering, and technology, only limited numbers have primary expertise in education that allows them to develop effective education products on their own.

The workshop summarized here was prompted by a number of changes both in NASA policy and in how the United States as a whole is changing the teaching of science in kindergarten through grade 12. The larger context of the workshop involves several significant events. These are the 2012 NRC report *A Framework for K-12 Science Education*[4] (generally referred to as "the Framework"), a set of K-12 science standards based upon the Framework known as the Next Generation Science Standards (NGSS), and the November 2014 release by NASA SMD of a Cooperative Agreement Notice (CAN) soliciting proposals that address NASA SMD's science education requirements.[5]

A Framework for K-12 Science Education

A Framework for K-12 Science Education (i.e., "the Framework"), released by the NRC in 2011, consists of the most up-to-date information on how students in grades K-12 should learn science (see Figure I.1). The development process of the Framework study consisted of a committee that included science education policy experts and researchers. Design teams in the following disciplines were utilized in the development process as well: engineering, Earth and space science, life science, and physical science. The Framework includes research on how students acquire knowledge of science in an effective manner, and it served as the basis for the NGSS, which were developed to provide an international benchmark for science education.[6]

---

[3] National Research Council, *NASA's Elementary and Secondary Education Program: Review and Critique,* The National Academies Press, Washington, D.C., 2008.

[4] National Research Council, *A Framework for K-12 Science Education: Practices, Crosscutting Concepts, and Core Ideas*, The National Academies Press, Washington, D.C., 2012.

[5] NASA, "A—Draft SMD Science Education Cooperative Agreement Notice," Solicitation Number NNH15ZDA002J, FedBizOpps.gov, posted November 6, 2014, http://www.fbo.gov.

[6] Next Generation Science Standards (NGSS), "Framework for K-12 Science Education," http://www.nextgenscience.org/framework-k-12-science-education, accessed January 15, 2015.

THE THREE DIMENSIONS OF THE FRAMEWORK

1 Scientific and Engineering Practices

1. Asking questions (for science) and defining problems (for engineering)
2. Developing and using models
3. Planning and carrying out investigations
4. Analyzing and interpreting data
5. Using mathematics and computational thinking
6. Constructing explanations (for science) and designing solutions (for engineering)
7. Engaging in argument from evidence
8. Obtaining, evaluating, and communicating information

2 Crosscutting Concepts

1. Patterns
2. Cause and effect: Mechanism and explanation
3. Scale, proportion, and quantity
4. Systems and system models
5. Energy and matter: Flows, cycles, and conservation
6. Structure and function
7. Stability and change

3 Disciplinary Core Ideas

Physical Sciences
PS1: Matter and its interactions
PS2: Motion and stability: Forces and interactions
PS3: Energy
PS4: Waves and their applications in technologies for information transfer

Life Sciences
LS1: From molecules to organisms: Structures and processes
LS2: Ecosystems: Interactions, energy, and dynamics
LS3: Heredity: Inheritance and variation of traits
LS4: Biological evolution: Unity and diversity

Earth and Space Sciences
ESS1: Earth's place in the universe
ESS2: Earth's systems
ESS3: Earth and human activity

Engineering, Technology, and Applications of Science
ETS1: Engineering design
ETS2: Links among engineering, technology, science, and society

FIGURE I.1 Framework for K-12 Science Education produced by the National Research Council. SOURCE: National Research Council, *A Framework for K-12 Science Education: Practices, Crosscutting Concepts, and Core Ideas,* The National Academies Press, Washington, D.C., 2012, p. 3.

NEXT GENERATION SCIENCE STANDARDS

The NGSS are a set of K-12 science standards developed through a state-led process to provide students with a benchmark for science education.[7] These standards are based on the NRC's Framework.[8]

The NGSS were produced due to the time gap in the development of guiding documents for state science education standards and the need to build interest among K-12 students in STEM disciplines. The standards are meant to better prepare high school students for college and the workforce with the objective of providing employers with the ability to hire individuals with strong science, critical thinking, and problem-solving skills.

Each NGSS consists of the following three dimensions: core ideas, science and engineering practices, and crosscutting concepts. Core ideas are meant to focus science curriculum and instruction on the most significant aspects of the discipline. Practices are applicable to both scientists and engineers; they describe the behavior of scientists as they build theories pertaining to the natural world and the practices of engineers as they build systems. Crosscutting concepts link different science domains, and examples include cause and effect, as well as energy and matter.[9] The focus of the standards is a progression of knowledge from grade to grade starting in kindergarten all the way through 12th grade. The standards emphasize engineering and technology, and they coordinate with the Common Core State Standards in mathematics as well as English language arts. The NGSS were released in April 2013 for adoption by states and continue to be implemented today.[10]

NASA's Cooperative Agreement Notice

NASA SMD's draft Science Education CAN issued in November 2014 sought comments from members of formal and informal education, and science research communities.[11] According to SMD, the directorate's vision for education is as follows:

> To share the story, the science, and the adventure of NASA's scientific explorations of our home planet, the solar system, and the universe beyond, through stimulating and informative activities and experiences created by experts, delivered effectively and efficiently to learners of many backgrounds via proven conduits, thus providing a return on the public's investment in NASA's scientific research.

The draft CAN was issued for a 30-day discussion period with a request for responses by mid-December 2014. NASA chose to use a cooperative agreement in lieu of a contract or grant, with the expectation that the agency would engage in substantial interaction with the parties that are selected.

A cooperative agreement occurs when there is a transfer of something of value to an entity, such as a municipality, state government, or private company, to be used for a public purpose. This legal agreement involves two parties: the federal government and another entity.[12] The goal of the CAN is to meet the following education objectives of NASA SMD: enable STEM education, improve science literacy in the United States, advance national education goals, and utilize partnerships to leverage science education. CAN awards are anticipated by September 2015, and NASA has the intention to select one or multiple science discipline teams(s).

---

[7] NGSS, "Science Education in the 21st Century—Why K-12 Science Standards Matter—and Why the Time Is Right to Develop Next Generation Science Standards," May 2012 Draft, http://www.nextgenscience.org/sites/ngss/files/Why%20K12%20Standards%20Matter%20-%20FINAL.pdf.

[8] NGSS, "Development Overview," http://www.nextgenscience.org/development-overview, accessed January 15, 2015.

[9] NGSS, "Three Dimensions," http://www.nextgenscience.org/three-dimensions, accessed January 15, 2015.

[10] NGSS, "Implementation," http://www.nextgenscience.org/implementation, accessed January 15, 2015.

[11] NASA, "A—Draft SMD Science Education Cooperative Agreement Notice," Solicitation Number NNH15ZDA002J, FedBizOpps.gov, posted November 6, 2014, http://www.fbo.gov.

[12] Kristen Erickson, NASA SMD, "NASA Science Mission Directorate Education Discussion with The National Academies Space Studies Board," presentation to the workshop, 2014.

1

Setting the Stage

The Workshop on Sharing the Adventure with the Student began with an introduction by NASA Science Mission Directorate's (SMD's) Kristen Erickson, a keynote address by John Mather, and a panel discussion among the four NASA SMD science education forum leads to provide the workshop participants with an understanding of the current state of NASA SMD's science education efforts and plans.

INTRODUCTION

Kristen Erickson, NASA Science Mission Directorate

Marc Allen, deputy associate administrator for research of NASA SMD, introduced the first speaker for the workshop, Kristen Erickson, who is SMD director for science engagement and partnerships. Allen explained that the administration's recent initiative to consolidate education within three agencies in the federal government prompted SMD to develop "a better and more efficient internal structure for our program, to align the education activities at the science theme level and less so at the individual mission level." Allen added that the associate administrator for SMD, John Grunsfeld, directed that they consolidate "education and communications within SMD's executive office to improve the integration of education across theme lines to better coordinate education with communications, and to increase the visibility of both within the agency."

Erickson explained why NASA is restructuring how SMD runs its education programs. NASA has developed a large and substantial education product portfolio, often as a result of individual space missions until recently being required to spend 1 percent of their budgets on education and public outreach. Erickson stated, "In the past 20 years on average, NASA Science has about 100 missions in development, in operation, or in extension. That supports about 100 or so different education teams funded in those missions." Erickson explained that such a large number of teams has made coordination difficult, and NASA wants to more effectively and efficiently connect with educators.

Erickson noted that one factor driving changes in how the agency approaches education is that the overall U.S. population has evolved over the past few decades. She also acknowledged that NASA is just one small part of the $600+ billion spent on education in the United States.

Erickson explained that the final text of the Cooperative Agreement Notice (CAN) will be released after January 1, 2015, and that NASA will hold a pre-proposal conference to answer any questions that potential proposers may have. "We use the cooperative agreement process so that you can propose what you're good at," Erickson explained. "You don't have to do the entire amount. If you are strong in one science discipline, please only propose to that. If you're currently on a mission in education and you know the scientists, you know the science; maybe you should lead a team."

Erickson outlined their objectives: "Enable STEM [science, technology, engineering, and mathematics] education. Improve U.S. scientific literacy. Advance national education goals. Leverage these efforts through partnerships," she said. "If these objectives sound familiar, it's because most of them were informed right here in these hallowed halls. All four of the recent NASA science decadal surveys addressed these objectives, the recommendations that were generated right here by the NRC [National Research Council]."

"We're also looking into a service that the Discovery class missions and Explorer class have adopted, which is a teaming website. This is a service that folks that are looking to team on our effort can go to see what skills are available or what skills they need," she added.

"What advice do we have for those that want to join us in this noble endeavor?" Erickson asked. "You are encouraged to think anew. If you are looking to propose to this CAN to promote your mission or your instrument, please do not. If you are looking to propose to this CAN to extend your [education and public outreach] grant for the next 3 years, kindly refrain. That's not what this endeavor is about. It's bigger. There's more at stake. If you really want to help us make a difference in this country's science education, propose to your strengths. You do not have to address every objective in the CAN."

She finished her talk by clarifying what needs to be done, encouraging proposers to use the cooperative agreement process to propose what they're good at. "Or, through this teaming website, you can reach out to others that may have like interests. We are not looking for one-offs. We're not looking to sprint. This is a long-term endeavor," she said.

LEARNING BY DOING

John Mather, NASA Goddard Space Flight Center

Growing Up a Scientist

John Mather of NASA Goddard Space Flight Center and recipient of the 2010 Nobel Prize in Physics gave the keynote address for the workshop (Figure 1.1). Mather explained his background and how it influenced him to become a scientist. "In my family we have seven teachers and three scientists," he explained. Mather said that he grew up on "a farm in northern New Jersey where dairy cattle research was being done," he remembered. "And there were also scientific labs with liquid nitrogen and Geiger counters and chemistry and scales and all kinds of stuff because people wanted to know how to get more and better milk from cows. It's more or less a solved problem these days, so this farm is no longer operating as a research farm. But it was a place for me to hear about science. I heard about cells and genetics when I was about six." It was called the Rutgers University Lusscroft Farm, which he referred to as the "site of early nerds in Sussex County, New Jersey."

Mather said that although he thinks this environment contributed to him becoming a scientist, he also wonders if it is somehow hereditary, considering all the teachers and scientists in his family. He went to public schools but was often sick, and so he had time to read and think at home. There was a bookmobile where he lived, and he read every science book that he could. "I always knew that I was a little different from the other kids. I knew that I didn't know why."

"I had lots of toys. I think this is something not every kid gets access to. Little mechanical toys, erector sets, chemistry sets, lenses to make telescopes, all those things I had when I was small." Being on a farm also meant that he was outdoors a lot, which gave him the opportunity to explore. He was exposed to geology and fossils and built forts and dams. Because his family was near New York City, he could go to museums such as the American Museum of Natural History, but also art and archeology museums. "Around the house there were hammers and drills and saws and nails, and all kinds of things that you can hurt yourself with. And I think that's pretty important for kids to have access to such things. Adults [should have access] too, to tell you the truth. I'm still growing up.

FIGURE 1.1 John Mather speaking at the workshop. SOURCE: Harrison Dreves, NRC.

I still enjoy looking around outdoors, seeing the geology of the world, driving through the countryside and seeing the layers of rocks and thinking about the history of our Earth."

Mather participated in science fairs starting in the fourth grade. He also attended summer camps, National Science Foundation summer schools in mathematics at Assumption College and physics at Cornell University. He eventually attended Swarthmore College and University of California, Berkeley.

Emphasizing the value of learning by doing, Mather said, "Nobody told me how to do my job when I got to NASA. They just said, *Here, try it*. And, *Go to these meetings, and people will help you, and we'll all think together.* And, golly, didn't it work! I don't think anybody could write you a book about how to be a project scientist or anything else that I've done in life."

Creating the Framework

Mather noted that several years ago he was asked to work on the NRC committee charged with developing a framework for K-12 science education,[1] which he said was an eye-opening experience for him. According to Mather, education standards have both good and bad aspects. The good is that they can indicate areas for improvement and help to celebrate learning. But if badly implemented, Mather said, then they can also mislead on how to improve, force "teaching to the test," "destroy all genuine curiosity," and discourage students by comparison

[1] National Research Council, *A Framework for K-12 Science Education: Practices, Crosscutting Concepts, and Core Ideas*, The National Academies Press, Washington, D.C., 2012.

to others. They can also "replace mastery with memory"—short-term memory. Finally, he stated, they can "waste our time and money and bore everybody to death."

Mather expressed dismay that most state science standards do not encourage hands-on learning. Students learn by doing. He joked about having standards of cooking. Students could memorize the cookbook and take a quiz. Or they could eat the cookbook. Neither would make sense. Instead, like science, the best way to learn cooking was to actually do it. "Cook something real and eat it! And try again tomorrow," Mather suggested. "If we had standards of bicycling . . . we couldn't really learn to ride a bike by reading a book. Instead, get a bike and try it out."

Mather said that students need to be encouraged to have a passion for learning about everything, both in and out of school and among all ages. Students need to have curiosity about the human-built world and the natural world. Mather would also like to see a joy of discovery and pleasure of accomplishment. All of this would help with preparation for adulthood, employment, community engagement, and parenthood. "Joy [of learning] is so important. Isn't it what propels students? I think we would all like kids to come out of school saying, 'Wasn't that fun? I'm glad I went. I want to keep on learning,'" Mather said.

Enchantment! Adventure! NASA!

As to the role that NASA could play in science education, Mather proposed that it could be summed up as "Enchantment! Adventure! NASA!" NASA can both engage and share the news. But Mather also noted that it is the risks that NASA takes that make the agency exciting. Mather said, "The risk makes it important. We risk a lot in our NASA programs, and once in a while people die. But the public pays attention partly because of that, believe it or not."

Mather said that there are many ways that the agency can provide the necessary engagement. They include team competitions at robotics and science fairs. "Life is a team sport from beginning to end," Mather said. NASA can also provide storytelling, tying science to daily life and culture. NASA can do inspiring things. "That's our job," Mather added. In addition, the agency can share the news via traditional media, social media, personal contacts, and visits. NASA can also partner scientists with educators to create real-world STEM materials and experiences for students. Again, he stressed that people learn by doing, even in reading and math. He mentioned Dean Kamen and FIRST Robotics and David Christian and Bill Gates's "Big History Project" as examples of recent efforts to engage students in hands-on activities in engineering and science.

Mather is currently the senior project scientist for the James Webb Space Telescope mission, and, as an example, he also showed some photos provided by the Space Telescope Science Institute of students building models of the telescope. Some of these models, he noted, are edible (Figure 1.2).

In response to a question from the audience, Mather considered a question that underpinned much of the discussion during the 2-day workshop: How can NASA support science education when there are only a few thousand scientists and engineers at NASA but 50 million students? It is not possible to directly reach all of those students, Mather noted, and teachers are the access. The key is giving teachers the opportunity to do science themselves. They need to experience the scientific process so that they can communicate it.

A member of the audience pointed out that there are many impediments to teaching science, including people who have an anti-science agenda. But Mather responded that anti-science agendas have existed since the Renaissance. Both Galileo and Magellan faced opposition, he noted.

Mather added that learning by doing is the opposite of learning from a book. He suggested that maybe the standard for learning should be that you should *do* something rather than describe it.

Mather thinks that NASA provides the way in for students. He noted all of the things that make NASA attractive to students. The agency makes materials based on science, engineering, and math. "People want to be like us," he concluded.

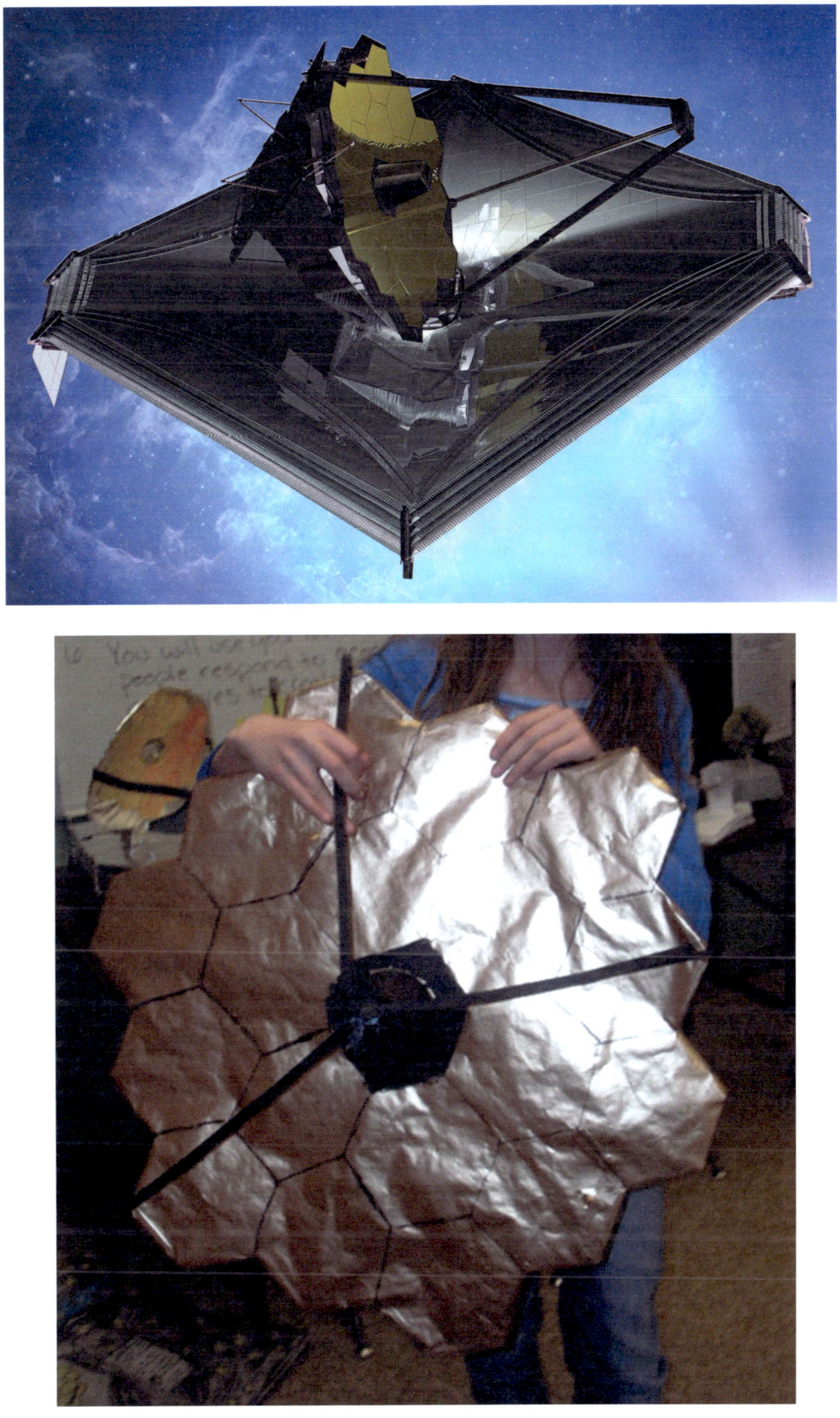

FIGURE 1.2 *Upper:* Artist impression of the James Webb Space Telescope. *Lower:* Student model of the telescope's hexagonal mirrors. SOURCE: *Upper:* Artist conception courtesy of Northrop Grumman Corporation; *Lower:* Courtesy of the Space Telescope Science Institute.

PANEL DISCUSSION WITH NASA SMD'S EDUCATION FORUM LEADS

The first panel of the workshop introduced the breadth of NASA SMD's educational activities. The panel consisted of four representatives of different education and science institutions who also lead NASA education forums:[2]

- Heliophysics Forum—Laura Peticolas, University of California, Berkeley
- Earth Science Forum—Theresa Schwerin, Institute for Global Environmental Strategies
- Planetary Science Forum—Stephanie Shipp, Lunar and Planetary Institute
- Astrophysics Forum—Denise Smith, Space Telescope Science Institute

Science and Science Education Go Hand-in-Hand

Denise Smith of the Space Telescope Science Institute noted that NASA is tackling fundamental questions in science. Materials from the Hubble Space Telescope are being extensively used, and scientists at the Space Telescope Science Institute are working with educators. "As we've already heard this morning, two of the ingredients that are fundamental and central to every student's [development]," she said, "are the educators that they encounter and their family—the people that shape them along the way. For that reason, NASA's Science Mission Directorate and the programs discussed here today work very closely with those educators and those families." Smith added, "We partner with a variety of formal and informal education organizations that are also working with and supporting those education and family audiences. And, as heard in today's introductory remarks, NASA has that ability to inspire and provide exposure to STEM with amazing role models and amazing science content."

Smith explained that SMD's science covers a range of disciplines. "The Science Mission Directorate spans the disciplines from Earth science to heliophysics to planetary science to astrophysics. The Earth and space science missions are tackling those fundamental questions: How does the universe work, where did we come from, are we alone?"

Smith continued that because of its breadth, SMD can have influence on different areas of science education. "The search for life and the conditions for life, what we learned by life on Earth, how the Sun interacts with the Earth and influences life on Earth, and the search for life in the solar system and beyond—all those spark fundamental curiosity in our students and make them want to explore," Smith told the audience. "SMD has been fundamentally working to share that science, to share that story, that adventure, with students, educators, and the public."

Smith said that "scientists bring unique expertise to the table, as do educators. The scientists know the science, they know the STEM career paths, and they work with the data. The educators are critical in the process of working with the scientists to translate the science to the classroom."

A Nationwide, Coordinated Community of Practice

Laura Peticolas of the University of California, Berkeley, explained that it is important to reduce duplication and to share best practices. She said that NASA needs to align its products to national education standards. "In the last 5 years or so, we have been working to reduce duplication across the nation. We now have a good sense of who is developing, for example, an activity on magnetism so that we're not developing 20 different magnetospheric activities. We have solid activities on comets. No one needs to develop any more comet lessons at this point."

Peticolas added that "we do make sure that our communities align their work to national education standards so that when the principal walks into the teacher's room and asks what they're teaching, they can address at least those standards, and often the local standards, which, of course, are more important for many of those teachers and school districts."

[2] NASA, "Science Education and Public Outreach Forums," last updated February 6, 2013, http://science.nasa.gov/researchers/education-public-outreach/science-education-and-public-outreach-forums/.

Putting Research into Practice: Curriculum Support Resources and Professional Development

Stephanie Shipp of the Lunar and Planetary Institute discussed the current state of practice for curriculum support resources and for professional development:

> Curriculum support brings together, again, those teams of education specialists and SMD scientists to create materials that augment classroom learning, not specifically to design full curriculum but to find places where NASA science intersects with curriculum learning and where we can enhance and support learning by the students and by the teachers. Often this is done in partnership with educational groups, such as Lawrence Hall of Science. [Curriculum support] brings together many of the missions and programs and individuals across the community so they can collaborate and focus on the larger questions, and, as Kristin [Erickson] said in her opening talk, not necessarily focus on specific missions or instruments.
>
> We always involve the students and the teachers in the development of curriculum material. Everything is driven by needs assessments in the educational literature, but also we have educators on the development team helping to make sure that the right decisions are made in the way things are presented.

Shipp also referred to another subject that would be addressed later in the workshop, the role of evaluation. She said,

> Evaluation is woven throughout from the beginning, the needs assessment, through pilot testing, field testing, and impact assessment at the close. Scientists on the team help ensure scientific accuracy. We also make sure that evaluations are aligned with educational pedagogies so that we meet the teachers' needs in the classroom, and make sure that [the evaluations] are tied to the National Science Education Standards. Once the materials are developed, they go through a NASA product review, then a SMD product review—where a panel of experts reviews the materials for relevance to NASA, scientific and educational accuracy, and alignment to the national standards.
>
> After materials are created, they then go through further review, and once they pass that review, they are disseminated through our partners and our educational networks. They are also disseminated on NASA Wavelength, which is SMD's digital library of products that have passed review. There are over 2,000 products in SMD's portfolio of reviewed materials [on Wavelength]. (See Box 1.1.)

Shipp repeated John Mather's comment about the small number of engineers compared to the large number of students. The goal is to "train the trainers," she said. "We only have a limited number of scientists; we only have a limited number of engineers. They all have day jobs, they can't go into every single classroom, but this is one method that may allow us to leverage those scientists and engineers in teams with the educators to train a core of master teachers and to build capacity and multiply the effect."

Collaborations, Partnerships and Evaluation

Theresa Schwerin of the Institute for Global Environmental Strategies discussed the importance of partnerships. "Partnerships have been developed over many years," Schwerin explained,

> These are very deep, very impactful partnerships, and they range over a broad spectrum of audiences from the Girl Scouts to groups serving very specific audiences such as the National Federation of the Blind. One partnership in particular is Earth to Sky, an interagency program with NASA, National Park Service, and the U.S. Fish and Wildlife Service. NASA Earth scientists and climate scientists work with interpreters and with NASA educators to help those who are on the frontlines with the public and with schools talking about climate science. This partnership has reached over 4 million visitors.

Schwerin reiterated that evaluation is a key part of all educational activities. She referred to an evaluation that compared students' performance with the standard curriculum to their performance after using NASA's Global Precipitation Measurement (GPM)-enhanced curriculum. She said testing indicated that both groups tested equally

BOX 1.1
NASA's Wavelength Website for Educators

NASA Wavelength[1] is a digital collection comprised of resources for educators, ranging from elementary to college level, focused on Earth and space science (Figure 1.1.1.). When speakers referred to the Wavelength during the workshop, they generally meant the website that hosts the collection, although the term refers to the entire collection. Wavelength was developed by a team from the following entities: the Adler Planetarium, the Institute for Global Environmental Strategies, the Lawrence Hall of Science at the University of California, Berkeley, and the Space Sciences Laboratory, some of whom participated in this workshop. Science Mission Directorate (SMD) funding resulted in the development of these resources, which undergo a peer-review process. Peer reviews are sponsored by SMD and include both scientists and educators who ensure the high quality of NASA education products and give feedback to the developers of the products.[2] Resources on the NASA Wavelength website include data and images as well as "strandmaps," which illustrate concept connections and demonstrate concept building across grade levels.[3,4]

---

[1] NASA Wavelength, "About," http://nasawavelength.org/about, accessed January 15, 2015.

[2] Institute for Global Environmental Strategies, "NASA Earth and Space Science Education Product Review," https://www.strategies.org/education/nasa-product-review/, accessed January 15, 2015.

[3] NASA Wavelength, "NASA Science Data and Images," http://nasawavelength.org/data-and-images, accessed January 15, 2015.

[4] NASA Wavelength, "Strandmaps," http://nasawavelength.org/strandmaps, accessed January 15, 2015.

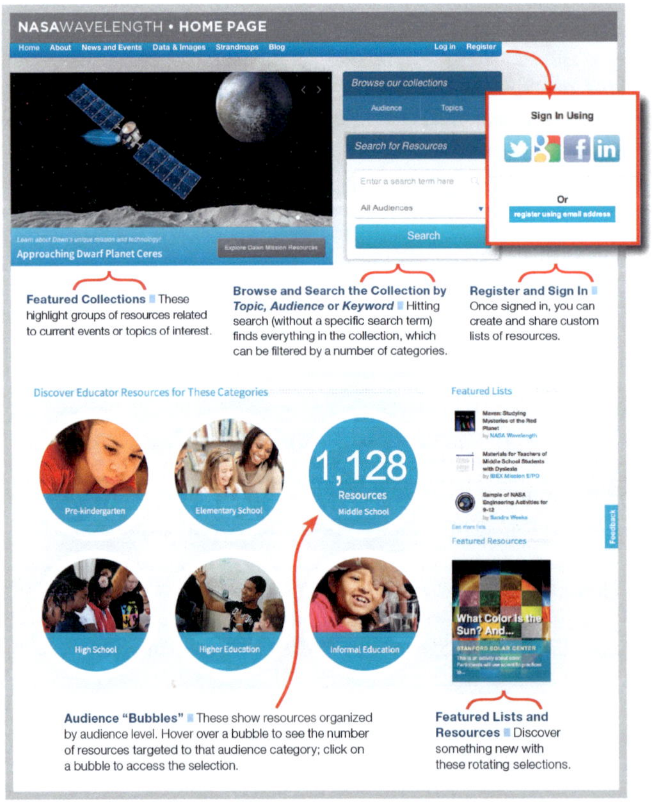

FIGURE 1.1.1 NASA's Wavelength website for educators makes available a substantial amount of resources. SOURCE: Courtesy of NASA and the Institute for Global Environmental Strategies.

immediately afterwards, but 6 months later, the GPM group members had better retention of what they had learned. This evaluation showed that learning science in context is more meaningful, Schwerin stated.

The audience members asked a number of questions. One asked, What is the next great challenge? The panelists responded that the coming year is a transition year. There is going to be a big change in how science standards are taught nationally as the NGSS are adopted by more states.

Another member of the audience noted that NASA's Wavelength site is not on the agency's homepage. There was some discussion among the panelists about the site's accessibility. Some commenters suggested that Wavelength is easy to find with a search engine like Google, and most educators start most of their searches that way rather than going to NASA's homepage. This issue of access to NASA's materials via the Internet came up several times during the remainder of the workshop. Additionally, an audience member mentioned that the Wavelength site has a heterogeneous selection of materials, commenting that he searched for "Keplerian orbits" as an example but found nothing. Continuing to search, he looked for "orbits," and this search returned only pointers to a DVD by the Astronomical Society of the Pacific. The audience member said, "Teachers need support in what they are required to teach, and we need to be able to reach out massively, even more [than what is on the Wavelength website]."

2

A New Vision for K-12 Science and Engineering Education and NASA SMD Education

Moderator	Brett Moulding, Utah Partnership for Effective Science Teaching and Learning, and Organizing Committee Co-Chair
Speaker	Stephen Pruitt, Achieve
Panelists	Maya Garcia, Office of the State Superintendent of Education, Government of the District of Columbia
	John Ristvey, University Corporation for Atmospheric Research
	Holly Ryer, Space Telescope Science Institute
	Sam Shaw, South Dakota Department of Education

NEXT GENERATION SCIENCE STANDARDS AND THE FRAMEWORK FOR K-12 SCIENCE EDUCATION

Stephen Pruitt, Achieve

Teaching the Underlying Concepts

Stephen Pruitt, senior vice president for content, research and development at Achieve, gave the address to begin Session 1. Pruitt, who led the development of the Next Generation Science Standards (NGSS),[1] explained that he started his career as a teacher.

> I remember vividly discussing the Mars mission and having two of the football players explaining how the need to have the weight room on the ship as they travel to Mars was important to not lose muscle in space. And I said, 'How are you going to have a weight room when weight has something to do with gravity?' The students said, *We've got a plan for that, we're going to close the door really fast before we take off and trap it in.* At that point in my career, (a) I realized I had not done a really good job of teaching [gravity], but, (b) I also started to realize that teaching is about helping kids understand scientific principles and how to apply them to phenomena and be able to explain multiple things with some very specific and strategic concepts.

[1] For more information, see the Next Generation Science Standards (NGSS) website at http://www.nextgenscience.org/, accessed January 15, 2015.

Pruitt eventually moved from teaching to education administration. There, he learned that the problems of teaching were not simply isolated. They were part of a broader societal struggle with explaining science and science education.

Pruitt emphasized that the National Science Education Standards (NSES),[2] like all good scientific theories, are built on the theory that preceded them. He said that the NSES was a great document when it was released in the 1990s, but many things have happened since the 1990s—texting, DVDs, and reclassifying Pluto as a dwarf planet, for example. When Pluto was changed to a dwarf planet, Pruitt spent an hour on the phone with a reporter from the *Atlanta Journal Constitution*, who kept asking when [Pruitt] was going to send a personal letter or e-mail to every teacher in the State of Georgia explaining that Pluto would no longer be tested. Pruitt said that he grew frustrated with the reporter.

> I kept explaining, "That's not the point. We don't test the planets."
>
> The reporter said, "Wait, you mean you don't test the order of the planets?"
>
> "No, we don't test them. We talk about what planets are, what the driving force in the solar system is, and talk about how excited the town was to see this kind of debate going on and how science really works."
>
> And the reporter said, "So let me get this straight, you don't think that every teacher needs to know you're not testing?"
>
> I replied, "We are not testing it *now*, we are not going to change because its status changed."
>
> I knew that I was going to end up in magazines across the country. I was going to be on bulletin boards in every classroom.
>
> At the end, as we were hanging up, the reporter said, "Well, Dr. Pruitt, do you think that kids are still going to use solar system models for science fair projects?"
>
> And flippantly I said, "Yes, but now they only need eight balls instead of nine." That was the only quote that got into the article. Next day, there was a political cartoon that had a picture of the CEO of Styrofoam Balls Inc. saying, "*Bad news, we predict a one-ninth decrease in our sales over the next school science fair season.*"

Pruitt explained that not only is this a good laugh, but also it is important because the reporter did not understand context. He did not understand that it is not just about memorizing planets, but it is about actually understanding what makes the solar system work.

Pruitt acknowledged that the scientific concepts are not easy. He said, "For instance, in any school in America, you could find a model. About 80% of them are edible—Jell-O molds of a cell with a big cookie with green icing representing mitochondria. Are those really models? I would say no." Pruitt continued that students have to identify the components of a system and then understand the connection between those components and what is it about the components that build to an explanation. Finally, the student has to identify the relationship between those components and the phenomenon that the student is actually studying. In other words, can this model actually be used to explain the evidence, or can it predict future phenomena? Pruitt said, "*That* is a model. Jell-O molds will not quite get there. In fact, I've jokingly said that one of the definitions for how you know if something is a model or not is: can you eat it? If you can, maybe it is not a model. Maybe it is a *representation*."

Pruitt said that science classes need to bring back the wonder of science. He does not want science to be a "word wall" with a bunch of science phrases on it that get crossed out as the teacher teaches them to the students.

Pruitt referenced the Framework's[3] core ideas (see Figure I.1):

> These core ideas, when you look at them in the framework and the NGSS, it is really about what has explanatory power. When you think about a meteorite hitting the Moon and shaping the Moon's surface, is it really about the meteorite? Is it really about the Moon's surface? Or is it about what it can tell us about the history of the Moon?

---

[2] National Research Council, *National Science Education Standards*, National Academy Press, Washington, D.C., 1996.

[3] National Research Council, *A Framework for K-12 Science Education: Practices, Crosscutting Concepts, and Core Ideas*, The National Academies Press, Washington, D.C., 2012.

> Meaning, maybe I need to understand that the bigger the mass, the more force is exerted when it hit. Maybe I understand more about the acceleration and the velocity of it, not just simply [that] it hit.

Pruitt also referred to crosscutting concepts. "They may be some of the most powerful things out there, and they may also be the forgotten dimension in a lot of places," he admitted. "These are the things that really bring the sciences together. They provide a cognitive structure for kids to be able to think about science. It's the fact that when we talk about energy and biology, it's the same as the energy we talked about in chemistry, the same energy we talk about in physics," he said.

Pruitt said that one promising development in recent years is that engineering is now part of the conversation on science. Until recently, engineering was often treated as part of a career track, not within the same realm as science and mathematics.

However, science education still remains trapped in other old-style concepts, Pruitt said. For example, every science textbook starts with the scientific method. But the scientific method is not how science actually works. Instead, it is a much more convoluted and iterative process. According to Pruitt, an important objective is to teach students to appreciate the practice of science, not the "scientific method." (See Figure I.1 for the list of scientific and engineering practices in the Framework.)

Aligning with the NGSS

Pruitt discussed how the NGSS were written not to guide the states, but to push the system to adopt new standards. He described a vision for instruction where the teachers are not "standing and telling"; instead, the children are learning and performing to demonstrate an understanding of doing science.

What are the implications of the NGSS for NASA? According to Pruitt, NASA will need to change its materials to better align. But alignment is not simply "checking the box." Alignment is making programs and lessons more like the Framework and the NGSS. "We are going to see a massive change in what is expected in classrooms. Part of what has happen is to deliver quality instructional materials. I believe NASA can do that. I believe NASA gives a real opportunity to bring back the wonder," Pruitt concluded.

PANEL DISCUSSION

Maya Garcia, science, technology, engineering, and mathematics (STEM) specialist for the Office of the State Superintendent of Education for the Government of the District of Columbia; John Ristvey, director of the University Corporation for Atmospheric Research; Holly Ryer, an education specialist at the Space Telescope Science Institute (STScI); and Sam Shaw, team leader of the Division of Learning and Instruction for the South Dakota Department of Education, joined Brett Moulding and Steven Pruitt for the panel discussion (Figure 2.1).

The organizing committee developed the following guiding questions to provide focus to the panel discussion:

- Present an overview of the NGSS and the role of NASA in supporting science and engineering education.
- How can/does NASA interact effectively with the education system to support K-12 science and engineering education?
- What opportunities does NASA Science Mission Directorate (SMD) have to better support the new vision described in the National Research Council's *A Framework for K-12 Science Education*?
- How can/does NASA integrate the science and engineering talent of NASA SMD into the SMD education programs?

Ryer explained that STScI has always developed materials based on educational needs. She said that the STScI's work is attractive to students because it not only addresses content, but it also has students using actual solar system data and performing statistical analyses with those data. Students organize, interpret, and draw conclusions about relationships between solar system objects. Rather than considering each solar system object in

FIGURE 2.1 Panel discussion for Session 1. SOURCE: Harrison Dreves, NRC.

isolation, students consider those objects as members of families. For example, Pluto is a member of a family of objects known as dwarf planets.

Ryer said that the NGSS focuses on science investigations and engineering and creating models of the real world. Similarly, NASA products can emphasize statistics, families, and groupings, but with real data that interest students.

Shaw explained that it is important for science teachers to understand how the work that NASA or a scientific organization does fits into their curriculum. "They [the content creators] think they're doing teachers a favor by saying 'This will fit kindergarten through 12th grade,'" Shaw said, "and what ends up happening is a volcano model, and it is an edible model."

Garcia explained that the Framework has two goals: all children will understand science well, and what students are taught will "act as a springboard" for some to eventually pursue scientific careers.

Ristvey stated that one of the values of integrating scientists and engineers into education is that they are practicing in a field that could be open to current students. They can help children understand what a career path in the sciences looks like. Scientists and engineers can make connections with project-based learning. NASA also provides real-life resources: men and women doing science and engineering using cutting-edge technology.

A member of the audience asked how the NGSS can be a springboard for STEM students and meet the needs of all children. Garcia answered that it is important to make learning accessible, no matter what the students starts with—if they are diverse groups, have special learning needs, or are performing above average. She noted that the three dimensions of the Framework allow teachers to recognize and assess students' understanding and then adapt and build learning around them.

An audience member asked how the Framework can help poor, underachieving schools and students. The panelists responded that the nation is making progress, but that first and foremost, champions for STEM education are needed. STEM needs to be more accessible as a content area in general. A panelist explained that when speaking to non-science people, it is important to put science in context. NASA can provide leadership for how science is a part of everyday life.

Another question from the audience was how to get teachers to be more "math-enabled." Pruitt said that one problem with past approaches was that they did not make clear connections between math and science. The math and science standards have to be coordinated. A member of the audience chimed in by saying, "Einstein had a famous quote that science without math is lame and math without science is blind."

In response to a question about assessments, Pruitt suggested that educators should not lead with a test. He said that it would be better to lead with building capacity for teachers to implement the NGSS and then build a test that matches what is going on in the classrooms.

3

Space Science Education Curriculum and Materials

Moderator Richard McCray, Visiting Scholar at the University of California, Berkeley
Speaker Edna DeVore, SETI Institute
Panelists Beth Johnston, Principal at Endeavour Elementary School
Mordecai Mac Low, American Museum of Natural History
Cassandra Soeffing, Institute for Global Environmental Strategies
Belinda Wilkes, Chandra X-Ray Center

BRINGING SPACE DOWN TO EARTH AND INTO THE CLASSROOM

Edna DeVore, SETI Institute

Edna DeVore of the SETI Institute gave the address to begin Session 2. She reminded the audience of the NASA Science Mission Directorate vision for education:

> To share the story, science, and the adventure of NASA's scientific explorations of our home planet, the solar system and the universe and beyond through stimulating and informative activities and experiences created by experts, delivered effectively and efficiently to learners of many backgrounds via proven conduits, thus providing a return on the public's investment in NASA's scientific research.

DeVore then explained that the federal government's National Science and Technology Council created a 5-year strategic plan that established five priority science, technology, engineering, and mathematics (STEM) education investment areas:[1]

- Improve STEM instruction;
- Increase and sustain youth and public engagement in STEM;
- Enhance STEM experiences for undergraduate students;

[1] Committee on STEM Education, *Federal Science, Technology, Engineering, and Mathematics (STEM) Education 5 Year Strategic Plan*, National Science and Technology Council, Executive Office of the President, May 2013, http://www.whitehouse.gov/sites/default/files/microsites/ostp/stem_stratplan_2013.pdf.

- Better serve groups historically underrepresented in STEM fields; and
- Design graduate education for tomorrow's STEM workforce.

Each of these investment areas has specific goals. For instance, improving STEM instruction has the goal of "preparing 100,000 excellent new K-12 STEM teachers by 2020 and to support the existing STEM teacher workforce." NASA's involvement in that goal includes projects such as the Airborne Astronomy Ambassadors.

DeVore stated that there are 67,000 elementary schools and 25,000 high schools with a total of approximately 50 million students. There are 1.7 million elementary school teachers in the United States, including 18,000 science teachers and 32,000 math teachers. There are 1.7 million high school teachers—209,000 science and 250,000 math.

NASA education programs have to "think nationally, yet serve state and local needs," she said. She added that sending a few scientists and engineers into classrooms cannot solve the problem, because it does not reach enough students. There are 60 times as many science teachers as there are scientists and engineers at NASA. DeVore said there is a need to match the scientists and their expertise to what the school requires, and scientists often fail to recognize that. She mentioned that she once received a phone call from a scientist who wanted to produce a lesson on solar seismology for third graders—the scientist did not understand what was suitable for third graders.

One of the ways to scale up, DeVore said, is to work through big partnerships, such as with publishers. An example of a partnership is the Full Option Science System (FOSS) Planetary Science middle school module. This is a kit-based course, with a hard copy teacher guide, student materials, plus web-based resources. Working with the publisher, Delta Education, the Kepler Observatory's education and public outreach program provided information for the newly revised course. DeVore said that the FOSS curriculum is used in all 50 states by more than 100,000 teachers and 2 million students and is in approximately 16 percent of the nation's school districts. It is adopted in 50 of the 100 largest urban school districts and reaches large populations of underserved students.

DeVore also noted that there is a digital divide based on household income: poorer students do not have access to computers or bandwidth. At schools, there are 3.1 students per computer. As an example, she noted, the U.S. Census indicated that 77.4 percent of white (non-Hispanic) households have Internet access, but only 61.3 percent of black and 66.7 percent of Hispanic households have Internet access. There is a significant divide based on income, with lower income households having significantly less Internet access than higher income households. DeVore said that the digital divide adds to the challenge of sharing NASA's stories.

DeVore concluded by saying that NASA education can make a difference, but it has to leverage partnerships to maximize reach and impact, and it has to "be attentive" to the standards, including the Framework, the NGSS, and state standards. "There will always be the need to have the big view but the local application," she said.

PANEL DISCUSSION

Beth Johnston, principal at Endeavour Elementary School, Kaysville, Utah; Mordecai Mac Low, curator and professor in the Department of Astrophysics at the American Museum of Natural History; Cassandra Soeffing, K-12 science education specialist at the Institute for Global Environmental Strategies; and Belinda Wilkes, director of the Chandra X-Ray Center, joined McCray and DeVore for the panel discussion.

The organizing committee developed the following guiding questions to provide focus to the panel discussion:

- How do the instructional strategies advocated for in the NASA education programs match the Vision for Science Education described in the *NRC Framework for K-12 Science Education*?
- How can NASA best encourage and support teachers to use NASA education resources in the classroom?
- What is the mechanism by which NASA education programs' instructional content material will be aligned to the Framework and the Next Generation Science Standards (NGSS)?
- How will NASA programs measure how well NASA Education and Public Outreach materials align to the NGSS?
- Information technology is changing the way science is done (data mining and simulations, for example)—What new possibilities does this development raise for the science classroom?

Wilkes was the first speaker. "I'm the director of the Chandra X-ray Center, so I'm here as a science provider, a source of content," she said. Wilkes believes that there is a bigger societal image problem that they all face. "We need to change the perception of STEM," she said. "Instead of being nerdy, it needs to be cool and fun. And we need to start that at a young age. It is not something that is scary; it is something that is around us all the time."

One possible way to do this is to give teachers hands-on opportunities to see how interesting and fun science can be, Wilkes said. "Get teachers interested in science, and then when they go to the classroom, they'll be more engaged and try, and learn. If the teachers are really engaged in teaching the science, they will impart that excitement to the students as well," she explained.

Johnston joined the discussion by saying that her school has a space theme throughout, even extending to its architecture. It is "a physical environment that breathes science," she explained. This environment affects the students' attitudes. "And they love space. Every one of them," she said. "They love science. In fact, our goals are always tied into that." One of the school's guiding philosophies is called "Mission MARS. And MARS stands for Math, Accountability, Reading, and Science," Johnston explained. "Every student knows our goals for the year. Every parent knows them. And every bit of resource that we have is funneled into those goals."

The school's work extends beyond simply classroom instruction. "We have a summer space camp. We have after-school robotics. We have an incredible 'month of space,'" Johnston said. "We have 'visions of the universe' panels that we share with other schools. We have a planetarium that is portable that our kids get to go inside." Every area of the school is named after a galaxy, Johnston added.

Johnston noted that Utah has the lowest per capita spending on students in the United States,[2] which forces her school to seek out partnerships with local businesses and other organizations. The school has grown substantially, from 450 to 1,130 students in only 4 years. Yet despite this substantial growth, it is a top performing school.

Soeffing then spoke. "I'm from Sioux Falls, South Dakota," she explained, which happens to be close to the Earth Resources, Observations and Science (EROS) Data Center. Soeffing said that she learned about the data center when she was a teacher and decided to get in touch with it, becoming part of an activity for teachers known as the Global Learning and Observations to Benefit the Environment (GLOBE) Program. "The GLOBE Program took us out to the EROS Data Center and showed us what Landsat data was like. I had a chance to learn about remote sensing and image processing. I brought it back to my classroom. I taught sixth grade science kids how to do remote sensing. They were totally amazed," she said.

But despite this opportunity, Soeffing's school faced many difficulties with keeping students engaged. The school has a high leave rate because many students return to the nearby Indian reservation. She noted that in particular, many girls in her school took naturally to math and science but then left the school and no longer had the ability to engage in the school's science programs.

Mac Low was the next speaker. "I'm a research astrophysicist mostly working on numerical simulations of star and planet formation. I'm also a professor in and one of the leaders of a Master of Arts and Teaching Program that we've started at the museum focusing on Earth and space science in New York State middle and high schools," Mac Low said. He finds that these two perspectives collide.

Mac Low said that it is difficult to bring teachers and scientists together. "NASA education is someplace where teachers meet scientists," he said, "and that collision is not always pretty."

Mac Low also criticized NASA education materials. "Providing an activity, which is the predominant thing I find on NASA websites, is not a lesson plan," he said. Teachers are pressed for time. "The impedance barrier between a working teacher, who has a 40-minute prep period to prepare the next several days of lessons, and the NASA portal is high. If I go to NASA Wavelength, it's a little bit better," he acknowledged.

Mac Low stressed the importance of intensive teacher education: "Teacher education is expensive because it involves a lot of hands-on experience. To get teachers to become excellent science teachers, they're going to need to do some science. That needs to be infused into teacher education programs." Mac Low said that they are trying to do that with their own program. "We're giving them science experience and science content, but that is expensive. Costs start to be like other professionals' training costs. Lawyers and doctors are intensively trained.

[2] U.S. Census Bureau, "2012 Census of Governments: Finance—Survey of School System Finances," 2012 Current Spending Infographic, http://www.census.gov/govs/school.

Teachers need to be intensively trained if we expect them to be able to work as highly qualified professionals, which is what I think our kids deserve," Mac Low said.

One audience member suggested that NASA has been producing a lot of educational materials, but it has been chaotic. It may take a teacher a long time to sort out the good from the bad. Another commenter suggested that NASA could take on interns who intend to be teachers.

Johnston said that she had asked the science teachers in her school district how NASA could better support them. The most common response was that they wanted to know "what NASA is doing today." They are overwhelmed with curriculum and really need a short lesson plan and materials. A participant from the audience responded that the "Astronomy Picture of the Day" and "This Week @NASA" already provide the information the teachers were requesting.

A member of the audience from the Space Telescope Science Institute noted that relationships take a long time to nurture. He explained that they have developed relationships with newspapers over a long period of time so that when they produce a press release, the newspapers trust its accuracy. He suggested that the same long-term relationship is required for universities, publishers, and other partners. He asked whether the changes, and the upheaval it can have with the relationships, make them nervous.

Mac Low replied, "It does make me nervous, because I think that the sorts of relationships that allow NASA to really best serve the education community—K-8 and community college, for example—those kinds of relationships take a long time to develop so that the individuals representing NASA content are trusted partners. It also takes a long time so that the people on the other end—the audience that NASA is aiming at—are aware of what is available. And there is a potential for an abrupt change. I am concerned that folks who have not worked on any of the NASA Science Mission Directorate education content over the past 20 years will put in a proposal for a radically different kind of education program, and what has been developed will be abandoned," he said.

An audience member then asked the panel, "How do your teachers find curricular resources that they would like to use—as an actual principal talking with your teachers?"

A member of the panel replied, "When we designed [the NASA Wavelength website], we did a lot of front-end work looking at user needs and how people search for and find resources and found most people use search engines. Part of the design of Wavelength is open to bring in traffic. And just under 40 percent of its traffic comes directly into specific resources via search engines like Google, bypassing www.nasa.gov."

An audience member commented about the idea of teachers being educated and trained at the level of doctors and attorneys. Teachers are being paid $30,000 to prepare students for $80,000 jobs, and as NASA designs learning environments and educator professional development, it is critical to keep the teachers' environment in mind and to raise this discussion to the national level, he said.

4

Collaboration Among NASA SMD and K-12 Districts, Schools, and Teachers

Moderator Mitchell Nathan, University of Wisconsin, Madison
Speaker Gordon Kingsley, Georgia Tech
Panelists Kathryn Flanagan, Space Telescope Science Institute
James Lochner, Universities Space Research Association (USRA)
Lora Bleacher, NASA Goddard Space Flight Center (GSFC)

BUILDING THE STEM PARTNERSHIP TOOLKIT: CHOOSING YOUR SPOTS CAREFULLY, MEASURING TWICE, AND FINDING YOUR SPANNER WHEN YOU NEED IT

Gordon Kingsley, Georgia Tech

Gordon Kingsley of Georgia Tech gave the address to begin Session 3. He started with some basic questions:

- What are institutional arrangements that provide effective platforms for facilitating successful collaborations?
- What are the barriers to accomplishing common goals across collaborating organizations? and
- How can these barriers be overcome?

"How do we govern ourselves?" he asked. "How are we going to organize ourselves collectively? And particularly, how are we going to do this around the purpose of discovery and education?"

One of the challenges facing educators and governments alike, according to Kingsley, is that the overall nature of how school programs are managed is changing. He said,

> We are in an era of contention between two visions of how we should govern ourselves. A vision that is the older one of a federally centric, public-interest-oriented form of governance, and the more current resurgence of an older tradition in the United States, of a state-centric, market-oriented strategy of governance. These two views are in contention with each other. NASA itself was born at the high watermark of this federal-centric, public-interest-oriented form of governance. Our teachers and our students are experiencing the market-oriented, state-centric view of governance. They're in the maelstrom of this debate.

> When you try to transfer knowledge, you are not only trying to do an extraordinarily difficult task, but also doing it in two competing systems of governance, where our teachers are experiencing one reality, and our research scientists are experiencing another. Bridging those worlds together is a very difficult act. And so partnership—the thing I like to study—is at the heart of this relationship.

Partnership, Kingsley stated, is viewed as a policy tool by governments and their agencies. In science, technology, engineering, and mathematics (STEM) education, partnership and collaboration is a preferred policy tool for transferring the benefits of public research investments to education institutions and students. But this creates tension between federalism and knowledge transfer. He noted that it can result in an escalation of collaboration rhetoric, with the rhetoric outpacing the reality.

However, there is a fundamental problem, Kingsley said. The term "partnership" is ambiguous, and people bring their own values into it. "Some versions [of partnerships] were entity-based: Create a team, a group, an organization, a not-for-profit, some enterprise that will carry forward; it will be the partnership." Kingsley added that other forms of partnership were venue based: "We will fix this school, and we are bringing all these resources together collectively to do great works."

Kingsley continued that other definitions of partnership were process-based: "We will network. We will get together, hold hands, and exchange knowledge; I will feel your pain; you will feel my pain. It will be great; and we will have a process!"

But there were still others, "usually the people with the cash," Kingsley said, who had a different view of partnerships: "This is a contract; it is agreement-based. You have your list of deliverables, the things you will do; I have the things I am supposed to do; this is how we will do it."

"Now, here's the kicker," Kingsley said, "These were all views that were held by *one federal agency*; in fact, *one team* of federal folks engaged in the partnership." The result was a lot of confusion. "The problem is not the ambiguity about partnership; the problem is the lack of transparency and the lack of articulation of the values that we bring to the partnership. In other words, you create all sorts of crazy expectations about, *well, this is what I expect these people to do*; but you don't actually communicate those expectations." Kingsley explained that a barrier to effective partnering is being transparent and clear. It is important to have a conversation between the critical partners to establish what each partner means by "collaboration" and "partnership."

Kingsley said that partnering is a very old policy tool and is a preferred tool for mission science agencies. Many other countries do not use partnerships to the extent that the United States does. He explained that particularly in parliamentary systems of government, they are a lot more directive. The Ministry of Education says, "this is what you will do," and you do it up and down the line. Why does the United States partner differently? Kingsley asked.

According to Kingsley, in the United States there is a basic tension between sovereign authority at the federal, state, and local levels and the authorized, appropriated mission to do knowledge transfer to the schools. He said,

> NSF, for example, back in its early days, was goaded on by the Department of Defense in the afterglow of Sputnik: What we need are national science standards and national mathematics standards. And you, NSF, fledgling agency, should go out and do that. The NSF director at the time looked out upon that vast wilderness of sovereign authorities—of school districts, state superintendents—and said, I'll take a pass; that looks like a little bit above my throw weight.

Kingsley echoed a theme mentioned by a number of earlier speakers about the disparity in the size of budgets available to NASA and the entire national education field. "In fact, if we took all the budgets combined of all the mission science agencies and pitted them up against K-12 education enterprises across the country, you all are going into a nuclear battle with a pea shooter," he said.

Kingsley explained that a final barrier to the business of partnering comes from unwillingness to engage the costs of partnering. He said that a lot of teachers are "partnered out." There are a number of reasons for the burnout. One is workflow interruptions. According to Kingsley, that creates a difficult dilemma for teachers. He speculated about what the teachers would be thinking after such a request:

> You are asking me to do non-routine tasks? Am I at a point in my work life where I am ready to take on that type of thing? If so, can I absorb the interruption? And more importantly, will you remember that you assigned me to this particular project to work on behalf of our school district and not punish me later on when I am not conforming to all of the performance standards that I'm supposed to perform at?

A second issue of the cost of partnering is mutuality in absorbing what a social scientist would call transaction costs. Kingsley explained that this requires that NASA, as the provider, think through how to get people to access and use its material in a powerful way. But he also thinks that NASA may do a better job than many: "Compared to many other federal agencies and many other state agencies, NASA is actually a little bit further down the road with education outreach officers, with attempts to do some outreach directly."

Kingsley said a lot more engineers are beginning to be involved in STEM education, and they are supporting the effective partnering process. He has seen that engineers take a strong view of risk management, and this view can be helpful, particularly in how engineers look at funding. Allocating funding evenly may not be the best approach; instead, more resources may need to be invested in one place than in another.

Another challenge is what Kingsley called "cultivation and career paths for boundary-spanners," those who are the "sacrificial angels" that help the partnership to function. If someone crosses a boundary in their work, will it be supported in their career path? The federal government is good at this, universities less so, and local education institutions are bad at this, Kingsley believes.

"Boundary-spanners are critical because they are the glue that is holding it together." They need support, Kingsley concluded.

POSTER SESSION

The moderator for the poster session was Mitchell Nathan of the University of Wisconsin, Madison. During the poster session, the participants, panelists, and audience members were encouraged to circulate among the posters and then be prepared to return to the discussion later with their observations about what they had learned (Figure 4.1). The participants were given the following guiding questions to consider during their discussions and were provided with pages to take notes on each question for each poster:

- What are institutional arrangements that provide effective platforms for facilitating successful collaborations?
- How are evidence-based models for successful collaborations or partnerships being communicated across NASA education programs?
- How are proven models or strategies for scaling up and sustaining collaborations and partnerships being used in the NASA education programs?
- What are the barriers to accomplishing common goals across collaborating organizations? How can these barriers be overcome?

Titles and presenters for the posters are listed below. Abstracts submitted by the poster presenters prior to the workshop can be found in Appendix B.

- THEMIS GEONS Magnetometer Program: Sustaining Teacher Engagement in NASA Science for Over a Decade
  —Nancy Ali, Space Sciences Laboratory
- Space Explorers Club and Heliophysics Educator Ambassadors: Growing District and Teacher Partner Relationships for Sustainable and Significant Impacts
  —Lindsay Bartolone, Southwest Research Institute
- Mars Education: Providing an Evidence-Based Model for Authentic, STEM-Practice-based Learning
  —Catherine Bowman, Raytheon
  —Michelle Viotti, Jet Propulsion Laboratory
  —Sheri Klug-Boonstra, Arizona State University

FIGURE 4.1 The workshop featured a poster session for participants to discuss various aspects of collaboration and then return to the main room for discussion. SOURCE: Harrison Dreves, NRC.

- Sharing the Adventure: Observation-Based and Data-Based Examples
  —Lin Hartung Chambers, NASA Langley Research Center
- Building Digital Age Resources Through Sustained Partnerships: MMS and ISTE
  —Troy Cline, NASA GSFC
- Best Practices from the Earth to Sky Interagency Partnership
  —Anita Davis, NASA GSFC
  —Ruth Paglierani, University of California, Berkeley
  —Sandy Spakoff, U.S. Fish and Wildlife Service
  —John Morris, National Park Service
- Strategic Partnerships: The Key to Sustainability and Reach for SMD Education
  —Bonnie Eisenhamer, Space Telescope Science Institute
- Empowering Educators to Engage with NASA Mission Science
  —Dorian Janney, NASA GSFC
- LRO's Lunar Workshops for Educators: A Proven Model for Exceptional Teacher Professional Development
  —Andrea Jones, NASA GSFC
  —Lora Bleacher, NASA GSFC
  —Sanlyn Buxner, Planetary Science Institute
  —Marti Canipe, University of Arizona

- *Explore!* Engaging Children in Space Science in Libraries and Other Out-of-school Programming
  — Keliann LaConte, Lunar and Planetary Science Institute
- NASA's Chandra X-ray Observatory Education, Public Engagement and Communications Program
  — Kathleen Lestition, Smithsonian Astrophysical Observatory, Chandra X-Ray Center
- Indigenous Education Institute: Collaboration with Integrity
  — Nancy C. Maryboy, Indigenous Education Institute
  — David Begay, Northern Arizona University
- Connecting GLOBE Students to Satellite Mission Partnerships
  — Tony Murphy, Global Learning and Observations to Benefit the Environment (GLOBE) Program
- NASA/IPAC Teacher Archive Research Program (NITARP): Evidence of a Successful Partnership Over a Decade
  — Luisa Rebull, NASA/IPAC Teacher Archive and Research Program
  — G.K. Squires, University of Hawaii
  — V. Gorjian, Jet Propulsion Laboratory
- NASA Astrobiology Institute: Embedding E/PO in the Place of Science to Maximize Collaboration, Partnerships, and Impact
  — Daniella Scalice, NASA Astrobiology Program

PANEL DISCUSSION

Following the poster session, the audience members were encouraged to discuss with panel members what they had observed. The guiding questions and focus for this panel were taken from the poster session. Mitchell Nathan and Gordon Kingsley joined Kathryn Flanagan, James Lochner, and Lora Bleacher at the panel table.

Lora Bleacher explained that she is the education and outreach lead for the Solar System Exploration Division at NASA GSFC. James Lochner runs education programs at the Universities Space Research Association (USRA). He previously worked in the Astrophysics Division at GSFC where he developed education programs and curriculum support materials, and at NASA headquarters where he ran grant programs. Kathryn Flanagan stated that she is the deputy director of the Space Telescope Science Institute, whose education and outreach program extends back decades. Previous speaker Gordon Kingsley of Georgia Tech explained that his current work involves looking at partnerships in knowledge transfer across engineering design teams and STEM group interactions engaged with the design teams.

Bleacher started the discussion by noting that Kingsley had covered best practices in his presentation and that she would share her own view of best practices. She said that as she walked around the poster session, she saw the importance of building trust and doing so through personal connections (e.g., site visits, in-person meetings). She emphasized the significance of having all participants buy in from the beginning and having a dedicated champion on either side of the partnership. She also noted the importance of the promotion of participants to leadership positions (i.e., people bringing leadership to the table).

Lochner explained that in walking around the poster session he often asked how a partnership got started, and often it was a matter of initial contacts and identification of needs. He noted that strong partnerships are engendered when there is a mutual identification of needs and the partners work together from the start. Kingsley pointed out that many in the audience are natural sociologists and there is no one best way in which collaborations come together.

An audience member commented that he agreed with Kingsley's comments in his presentation that the federal-state interface has a lot of friction. Some NASA educators have explained that the reason why they are focusing on creating activities rather than lesson plans is because they do not want to be seen as stepping on the prerogatives of the different states to determine their individual requirements for curriculum and lesson plans. The audience member acknowledged that this may indeed not be NASA's job.

Lochner stated that the cultural differences between scientists and educators can be managed when scientists understand the needs of the partner and each side is honest about its needs. At the Space Telescope Science Institute, Ph.D. outreach scientists are embedded in the team, and communication is integrated within the group. Bleacher stated that barriers can be addressed by helping scientists to not only share their scientific work, but also share personal details so that it is easier for teachers to relate to scientists as people.

Bleacher explained that evidence-based models are shared through opportunities such as the workshop and that constant communication is important. Lochner added that one of the things that the NASA education forums realized is how important it is to communicate success stories, and by communicating these success stories they were communicating best practices. He noted that internal professional development is important and there are many ways in which leaders in the [education] community helped to build up the strength of community.

One member of the audience said that he is concerned that the opportunities for professional development may be lost. "I fear, in the new realm of this funding model of SMD," that some things are no longer happening. "The [NASA SMD Education] forums have fostered retreats over the last several years. And our organization has hosted an annual [education and public outreach] conference that has seen dwindling attendance, because there has been less of a push, it seems, from Headquarters for folks to travel to our meetings or to some of the retreats that the forums have hosted," he said. The audience member commented that he hopes there will still be value given to professional development workshops and other events as places to share best practices and to learn from each other.

Another member of the audience raised a question about the cultural differences between scientists and educators. "I was wondering if there have been any strategies that you have found effective in helping scientists and educators talk to each other in the same language so that scientists are not talking like they are doing peer review while educators are talking like they are supporting learners."

A panelist replied that they have seen some attempts at improving scientist communication at Georgia Tech. "The first attempt at it did not go quite so well, largely because it consisted of sitting the faculty members down and saying, 'Don't talk that way. You are arrogant and condescending and unpleasant.'" A later effort was more successful. The panelist described engagements where the teacher was teaching the professor. That allowed for value to be exchanged both ways, and this changed the nature of the conversation.

Lochner commented that there are many instances where cultures clash due to language differences that are not foreign in nature. He noted that different organizations have different ways of speaking (e.g., working with Native American communities, getting planetary scientists and biologists talking). He emphasized that the right approach is important.

An audience member commented about the discussion of engaging scientists in education. Engaging a scientist can be powerful if the right scientist is in the right situation. She pointed out that some are engaging and some are not, but the ones who are not may be good at other things, such as curriculum review or blog posts. Scientists can be prepared for the audience that they will engage, and this is powerful if done well by people who know both groups and how to engage effectively. Lochner pointed out that Edna DeVore's presentation charts sketched out ways in which scientists could be involved. He realized that he could help mediate the relationship between the scientist and the educator by knowing where training was necessary. Kathryn Flanagan stated that the Space Telescope Science Institute sometimes helps scientists with communication, and many in the science community are thinking about communication as part of their duty and as a privilege.

Another participant mentioned that GSFC tries to engage its postdocs early. "They tend to be excited and enthusiastic to begin with, a little more receptive sometimes to some constructive feedback on their presentation styles."

Bleacher replied that at GSFC there is a lunar workshop for educators, and it tries to get postdocs immediately engaged. GSFC has also recognized the importance of communication, and the center has a partnership with the Stony Brook University Center for Communicating Science. The Stony Brook University journalism department has also been involved to train others to talk about science ongoing at GSFC.

An audience member stated that the culture has shifted for astronomers engaged in outreach. She wanted to hear how the culture has evolved and where the panelists have seen that evolution. She asked the panelists to talk about the evolution of partnerships over the past 15 years. In response, Flanagan stated that 15 years ago, missions had to spend about 1 percent of their budget on outreach, and many scientists thought this was not career enhancing. However, things have completely changed. Young scientists at the Space Telescope Science Institute now *want* to do outreach. People did not know what to make of it in the beginning, but things are different now, he said.[1]

An audience member referenced the dedicated champion comment made previously and pointed out that one of most valuable champions is the Parent-Teacher Association (PTA). She questioned the panelists about any comments that they may have about working with the PTA or other interested organizations. Bleacher agreed that champions on the PTA could be helpful and that this is an intriguing point.

An audience member noted the power of champions that was discussed earlier and pointed out that the Standards and Framework represent an opportunity for change.[2] He asked the panelists what is the role of opportunity or challenge to a partnership?

Kingsley stated that windows of opportunity are taken seriously in his work. Timing and environments are important, and windows for effective change are fragile and fleeting and need to be seized when presented. Champions are also fragile because many pressures are applied to them, and the champion shelf life is short, he said.

Bleacher commented that she was encouraged by the poster sessions because they indicated the importance of making personal connections in partnerships. This included making sure that there is a dedicated champion on both sides of the partnership. She went on to say that many of the programs she saw during the poster session not only had shared goals, visions, and buy-in, but also people that were actually leading. The poster presenters were helping create the strategic vision, making the path forward, and looking for ways to increase sustainability of the program. She noticed that people were bringing not only shared resources to the table, but also their leadership.

Flanagan described the criteria that the Space Telescope Science Institute uses to choose partnerships that will be scalable and sustainable:

> Our specific criteria are: Does the institution complement the work we do? Does the partner serve underrepresented or underserved populations? Does the partner demonstrate sustainability, and can it carry on independently? Will there need to be a lot of hand holding after the first year or so, or will this be able to carry on? And finally, and it's important for us, does the partner collect follow-up data? Will they participate with us in giving us feedback and summative evaluation so we can know whether it works? And with those criteria as ideas, that is how we might embark on a particular partnership. Now, it turns out to be absolutely fundamental to the success of our scalability. We have three leveraging techniques we use that make a core team of four and a half people reach out to more than half the middle school students in the country and half the middle school teachers, one of which is the partners. One is that we do eventually have online access, and one is that we target our professional development for master teachers.

Nathan ended the session, commenting that he has noticed both the scientists' and the educators' points of view have changed. The scientists now see the educator professionals as being their partners and as important to bringing mission results to the public.

[1] W.R. Penuel, C.J. Harris, A.H. DeBarger, Implementing the Next Generation Science Standards, *Phi Delta Kappan* 96(6):45-49.

[2] See the "Introduction and Background" chapter for a discussion of the Next Generation Science Standards and the Framework for K-12 Science Education.

5

Supporting Science and Engineering Teachers Through Professional Development

Moderator Albert Byers, National Science Teachers Association
Speaker William Penuel, University of Colorado, Boulder
Panelists Annette DeCharon, University of Maine
Sheri Klug-Boonstra, Arizona State University
Mariel Milano, Orange County Public Schools, Florida

PREPARING TEACHERS TO SUPPORT THREE-DIMENSIONAL SCIENCE LEARNING

William Penuel, University of Colorado, Boulder

William Penuel, University of Colorado, Boulder, gave the address to begin Session 4. Penuel explained that if NASA wants its education materials to be used, then it will have to form partnerships with schools. He also called for developing a new set of professional learning standards and professional development research. Penuel also addressed how educators are trained and adapt to new changes in curriculum.

Defining "3D Learning"

Penuel began by talking about one of today's challenges in professional development research—preparing teachers to support three-dimensional science and engineering learning as presented in *A Framework for K-12 Science Education* ("the Framework),[1] a consensus report that informed the Next Generation Science Standards.

The Framework, Penuel continued, summarizes what we now know about how children learn and discusses three-dimensional science learning, which comes from a rich, multi-decade appreciation of what scientists actually do in their work and how we prepare young people for that. He went on to ask how we prepare teachers, especially when they may not have had such an experience.

Penuel said that in the past, student expectations for science focused on content and process separately. Motivated by a growing sense of the need for greater coherence in science education, he said, the Framework emphasizes that students need to develop an integrated understanding of science and engineering, both as a body

[1] See the "Introduction and Background" chapter for a discussion of *A Framework for K-12 Science Education*.

of knowledge and a set of practices for developing new knowledge. In addition, he noted that students are expected to apply crosscutting concepts that unify science and engineering—concepts such as structure and function, cause and effect—to deepen their understanding of those core ideas.

According to Penuel, coherence is manifest in the developmental perspective taken in the Framework. The Framework orients educators to help children "continually build on and revise their knowledge and abilities, starting from their curiosity about what they see around them and their initial conceptions about how the world works." It is guided by the logic of a progressive view of learning, in which students' knowledge and skill become more sophisticated over time, in ways that are directly supported by instructional materials and the actions of teachers.

Penuel stated that realizing the vision of the Framework depends on aligning the components of an educational system with that vision. It depends on making the system horizontally coherent, in the sense that the curriculum-, instruction-, and assessment-related policies and practices are all aligned with the standards, target the same goals for learning, and work together to support students' development of the knowledge and understanding of science.

Professional development is also an important system component to align, he said, because the Framework makes significant learning demands on teachers.

Penuel said that it is known from the first generation of science standards implementation that professional development of an extended duration that is focused on content and that is close to practice is necessary to change classroom practice.

Teachers' Struggles with Changes

Three of the changes for teachers are what Penuel considers to be particularly "big asks" of teachers, because they represent big departures from most teachers' current practice.

He said, the first "big ask" is the demand to focus on a few disciplinary core ideas and help students make connections to crosscutting concepts. Students frequently encounter science today as small, disconnected facts; the assessments given to them that only require memorization of definitions of key terms or the recollection of facts reinforce this notion. So, to focus on just a handful of ideas, to know how to help students learn to use those ideas to explain and predict phenomena, and to find creative ways to ask students to make connections across vastly different areas of science, is a big shift to ask of teachers, Penuel said.

Penuel continued, a second "big ask" is to engage students in science and engineering practices. The eight practices represent the key "verbs" of science—how it is that scientists come to know and communicate what they know about the natural world. When students are engaged in these practices, they come to learn what it means to develop scientific knowledge and to think, speak, and act like scientists. But there is a big difference between having students pose questions to investigate and having students ask questions to clarify instructions for an assignment. And there is also a big difference between having students plan their own investigations—struggling with making a scientific protocol, following it, and discovering that it does not work—and having them use a protocol someone else developed for them. Finally, Penuel noted that there is a big difference between orally defending one's findings before peers and scientists and writing up a lab report with a format they have been given to follow. He said that the practices demand that students wrestle with the mess, the mangle of science, and not have it cleaned up for them.

Penuel continued that a third "big ask" of teachers is to help all students build a stronger identification with science and engineering. According to the Framework, teachers of science and engineering need to make reference to the rich variety of human stories, to the puzzles of the past and how they were solved, and to the issues of today that science and engineering face, connecting them to their "human roots." He said that a key principle of the Framework is that teaching must build on the interests and everyday experiences of young people in their families and communities, helping them see the connections between their everyday activities and the disciplinary ways of knowing the world. At present, we know that in many classrooms—particularly beginning in middle school—interest in science and engineering declines, and so we have much to do to realize this principle of the Framework.

During the past 3 years, Penuel and his colleagues have been leading professional development activities with teachers in a number of school districts large and small across the country. Some recurring things he has observed and heard from teachers is that integrating core ideas and practices is not easy. Many teachers see the eight practices, and they say that "this is the scientific method" or "this is just inquiry."

He said that many are also rightly skeptical about making shifts to instruction, when there are no available materials for them to use that embody three-dimensional science teaching and learning, and when state assessments for which they are accountable do not align yet to the Framework.

Penuel explained that a strategy of looking on the Internet for activities will not do either, because teachers are unlikely to find coherent sequences of instructional activity that build student understanding over time in the way outlined in the Framework. Curriculum materials provide important models for teachers to use to organize their own instruction. These struggles demand of teachers an engagement of the ideas in the Framework, and they demand of us—curriculum developers, professional development providers, and educational leaders—that we partner with them to develop the materials and assessments they need.

Penuel explained that it is possible to organize professional development opportunities to help accomplish this task.

Current Conditions

At present, Penuel said, few publishers or funders are willing or capable to make large investments in the kinds of coherent, year-long, and multi-year curriculum materials needed to embody the Framework. This means that the wide variety of materials that teachers need across K-12 education may not exist for some years to come.

He explained that science continues to compete for attention, too, with other subject areas, and this competition is not only for time in the classroom but also for teachers' professional development time. Science teachers in districts are regularly called upon to help their colleagues in other disciplines to implement major initiatives rather than build knowledge and resources that advance their own work in science.

Penuel said that another challenge is that there are so many different professional development providers at the local, state, and national levels—including NASA. Each competes for teachers' attention, and it will not be easy to ensure that providers all have a robust understanding of the Framework or access to resources that embody and make visible the vision, principles, and substance of the Framework.

Penuel said that he mentioned these challenges, not because he thinks they doom their efforts to failure, but rather because they are considerations when designing professional development. How might we prepare teachers, for example, for making teaching shifts in the Framework but develop materials with them at the same time, he asked. And what might be the role of standards in helping bring diverse providers into alignment with respect to professional development practices?

Strategy 1: Focus on the Framework

Penuel said that we know from research on the first generation of standards implementation that teachers are likely to interpret new standards in light of existing ones and their own ideas about the nature of scientific practice. He said that we also know that the sense teachers do make of standards is likely shaped by the ideas they share with their close colleagues. So teachers will need structured time for making sense of the Framework, as part of learning about it. Ideally, this reflection would take place within school-based teacher communities, where teachers' interactions with colleagues are concentrated.

This is particularly important, Penuel noted, for helping teachers see what is the same and what is different about the Framework, and for focusing on the three-dimensional image of science learning presented in the Framework.

Focusing on tools and processes that teachers can use in their schools addresses the reality that there may be limited time for professional development in external workshops. External providers can design these tools, introduce them to teachers, and periodically bring back leaders of professional communities together to discuss their use of the tools, both to gain support from colleagues and acquire new ideas to try out in their schools.

One of the strategies Penuel and his colleagues have employed to support this work is to ask teachers in small groups to read parts of the Framework—either a single practice, crosscutting concept, or disciplinary idea, or the components of a performance expectation in the Next Generation Science Standards (NGSS).[2] They ask them to

[2] See the "Introduction and Background" chapter for a discussion of the Next Generation Science Standards.

work in small groups to develop "evidence statements" for what they would expect to see students see or do who had mastered that core idea or exhibited skill in the practice. They also ask teachers to think about different levels of performance they might see, which draws their attention to the learning progressions outlined in the Framework.

Of course the work of breaking down the practices, core ideas, and crosscutting concepts of the Framework into some component parts could be done for the teachers, but the intellectual heavy lifting they ask teachers to do helps them to wrestle with the specifics of the Framework and to do so publicly with others, so that different interpretations can emerge and be discussed as part of a learning community.

Penuel and his colleagues also sometimes ask teachers to read brief summaries or parts of articles on student learning related to parts of the Framework. Especially where there is research to support hypothetical learning progressions, this helps teachers anticipate how students might engage with the ideas in the Framework.

Another sense-making activity is to ask teachers to compare the usefulness of different tasks for eliciting student's three-dimensional science proficiency as represented by a performance expectation of the NGSS. In this activity, they find or develop a range of tasks and ask teachers to rate the quality of evidence that each would generate about student learning. This particular activity helps teachers readily discern differences between tasks that elicit only facts from those that require students to also use science or engineering practices and make connections to crosscutting concepts.

Strategy 2: Co-Design Curriculum with Teachers

Penuel said that engaging teachers in the design and adaptation of curriculum materials is a form of "active learning" in which teachers can effectively explore the materials through practice, investigation, problem solving, and discussion.[3] Developing materials provides a way for teachers to match instructional materials to student needs.[4] Engaging teachers in design and adaptation of materials provides a way for them to learn about theories from research they can apply to practice.[5] He said that teachers who begin with strong models of curriculum materials provide students with higher-quality opportunities to learn.[6] Professional development that provides models for adapting materials and teaching effectively with them can produce greater student learning gains.[7]

In Denver, Penuel is working with a team of teachers to co-design a curriculum in biology for ninth grade. The team is comprised of local scientists from both the university and the parks and recreation department in the city, as well as experts in curriculum from the Biological Sciences Curriculum Study in Colorado Springs and learning scientists at the University of Colorado.

The team's process looks little like traditional curriculum development by experts, or the typical efforts of districts to develop their own materials. It is a hybrid process that blends sense-making about the Framework with collaborative design as a leading activity of professional development. The project is funded by the National Science Foundation, through its Cyberlearning Program.

Penuel said that the team kicked off its design work, which is ongoing, with a 5-day workshop. The workshop included opportunities for teachers to learn about the Framework and the NGSS, including engaging in the activities. It also included opportunities for teachers to work both in a large group to develop unit structures and in small groups to dive deep and do design work together. At set times, the whole group convened to review the evolving unit for its overall coherence.

---

[3] E. Banilower, D. Heck, and I. Weiss, Can professional development make the vision of the standards a reality? The impact of the National Science Foundation's Local Systemic Change Through Teacher Enhancement Initiative, *Journal of Research in Science Teaching* 44(3):375-395, 2007.

[4] D. Huffman, K. Thomas, and F. Lawrenz, Relationship between professional development, teachers' instructional practices, and the achievement of students in science and mathematics, *School Science and Mathematics* 103(8):378-387, 2003.

[5] H. Parke and C. Coble, Teachers designing curriculum as professional development: A model for transformational science teaching, *Journal of Research in Science Teaching* 34(8):773-789, 1997.

[6] W. Penuel and L. Gallagher, Preparing teachers to design instruction for deep understanding in middle school Earth science, *Journal of the Learning Sciences* 18(4):461-508, 2009.

[7] W. Penuel, L. Gallagher, and S. Moorthy, Preparing teachers to design sequences of instruction in Earth science: A comparison of three professional development programs, *American Educational Research Journal* 48(4):996-1025, 2011.

To support the work of development, the team is using a tool developed by Brian Reiser called a storyline diagram. The storyline diagram helps to frame an overall phenomenon and, in their case, an engineering design challenge for the unit. In their ecosystems unit, students will be exploring the phenomenon of how humans are disrupting ecosystems by planting trees. Their challenge is to help the local parks and recreation department to select trees to plant in their schoolyard that will increase biodiversity and maximize benefits to human beings and other organisms. Like all design challenges, this one will engage students in wrestling with the inevitable trade-offs in thinking about "benefits" to the environment, and also see ways that human disruption is not always problematic in ecosystems.

The storyline helps to keep the unit coherent by reminding everyone where they are headed, and what students can explain about the phenomenon that is new after each cycle of activity. It's organized into micro-cycles of question asking and answering, engaging in science and engineering practices that have to fit together, and each contributes to some aspect of helping students develop the understanding they need to solve the design challenge.

Strategy 3: Formative Assessment about Student Interest and Experience

Penuel said that it is important to elicit students' prior knowledge in order to develop scientific understandings of phenomena, but they rarely find out what interests, experiences, and practices from home and community they bring that might be used to connect student learning in ways that help them identify with the enterprise of science.

One approach, he noted, that could prove fruitful for classroom assessment is a strategy used in an elementary curriculum unit called "Micros and Me."[8] The unit aims to engage students in the practice of argumentation to learn about key ideas in microbiology. In contrast to many curriculum units, however, this example provides students with the opportunity to pursue investigations related to issues that are relevant to them. The researchers adapted a qualitative methodology from psychology, photo-elicitation, which is used to identify these issues. Research participants take photos that become the basis for interviews that elicit aspects of participants' everyday lives.[9] In Micros and Me, at the beginning of the unit, students take photos of things or activities they do to prevent disease and stay healthy. They share these photos in class as a way to bring personally relevant experiences into the classroom to launch the unit. Their documentation also helps launch a student-led investigation focused on students' own questions, which are refined as students encounter key ideas in microbiology.

In describing the curriculum, Tzou and Bell[10] do not call out the practice of self-documentation of students' personally relevant experiences as a form of assessment. At the same time, they note that a key function of self-documentation is to "elicit and make visible students' everyday expertise" relevant to the unit content.[11] Eliciting and making visible prior knowledge is an important aspect of assessment that is used to guide instruction. It holds promise as a way to identify diversity in the classroom in science that can be used to help students productively engage in science practices.[12]

Professional Learning Framework

Penuel concluded by describing an effort of a committee of the Council of State Science Supervisors that he and another researcher, Richard Audet, are supporting. This is an effort to develop a new set of professional learning standards for science education, aligned to the Framework, which also updates the professional development standards outlined in the National Science Education Standards. He said that one of the things that a set of standards

---

[8] C. Tzou, L. Bricker, and P. Bell, "Micros & Me: A Fifth-Grade Science Exploration into Personally and Culturally Consequential Microbiology," Curricula, Everyday Science and Technology Group, University of Washington, Seattle, Wash., 2007.

[9] M. Clark-Ibañez, Framing the social world through photo-elicitation interviews, *American Behavioral Scientist* 47(12):1507-1527, 2004.

[10] C. Tzou and P. Bell, "Micros and Me: Leveraging Home and Community Practices in Formal Science Instruction," pp. 1135-1143 in *ICLS '10 Proceedings of the 9th International Conference of the Learning Sciences*, Volume 1, International Society of the Learning Sciences, 2010, http://dl.acm.org/dl.cfm.

[11] Tzou and Bell, "Micros and Me," 2010.

[12] Clark-Ibañez, "Framing the social world through photo-elicitation interviews," 2004; Tzou and Bell, "Micros and Me," 2010; Tzou et al., "Micros & Me," 2007.

can do is to provide guidance to providers, educational leaders, and teachers regarding professional development. To craft coherence across diverse providers, Penuel said that it is also important to facilitate processes by which a network of activities and providers come together to discuss the standards and align their work to one another to develop their own learning community.

He said that the magnitude of the "lift" that must be made to realize the vision of the Framework for all students is large. It requires big shifts for teachers and for professional development providers. It requires changes of systems, which are already overburdened by competing reforms and stringent accountability requirements for teachers and schools. But to succeed, Penuel said, we need to start somewhere and start small, while still thinking systemically about what it is we are building. Penuel said that he sees NASA as one of a number of providers of professional development with the potential to contribute as a partner to efforts to build a more coherent science education system.

PANEL DISCUSSION

Sheri Klug-Boonstra, director of Mars Education at Arizona State University; Mariel Milano, director of digital curriculum and instructional design for Orange County Public Schools in Florida; and Annette DeCharon, senior marine education scientist at the University of Maine, joined Penuel and Byers for the panel discussion. The organizing committee developed the following guiding questions to provide focus to the panel discussion:

- How are standards for professional development used in NASA professional development programs?
- How do the mechanisms and programs by which NASA programs meet the needs of in-service teachers, and how does this differ from the ways NASA programs meet the needs of pre-service teachers?
- What are the most effective and widely used delivery models (e.g., online, train the trainers, professional learning communities, summer seminars, internships) for NASA professional development programs?
- What are example strategies for partnering scientists and educators?

Klug-Boonstra explained that she was classically trained as a science education teacher, and she wanted to find a way to engage students. She ran after-school programs and modeled professional development for elementary teachers. She also previously ran the NASA undergraduate internship program across all NASA centers and eventually returned to K-12.

Mariel Milano stated that she began her career teaching and also served as a science coach, curriculum developer, and a professional development facilitator. She is currently working on developing NGSS-based curriculum with Hofstra University. She also oversees the implementation and conversion of the Digital Learning program in her district. Milano explained that she was on the NGSS elementary and engineering writing team. She emphasized that she believes that NASA's role is uniquely positioned as a professional development agency in collaboration with districts.

Annette DeCharon explained that her background is in geology and oceanography. She worked at the Jet Propulsion Laboratory as a mission planner on Earth and space science missions, and she was the lead for education and public outreach for the TOPEX/Poseidon Mission. She currently does communication and public engagement for the NASA Aquarius Mission, which measures global sea surface salinity.

Moderator Albert Byers began the panel discussion by referencing Penuel's presentation and the three strategies that he provided. Byers said, "Professional development is ongoing. It's not a one in, done, one-shot opportunity . . . during the 1-week summer institutes only." He asked the panelists to provide their impressions on how these strategies might inform their current efforts or knowledge of NASA. He stated that the first strategy was to simply focus on the framework. Teachers would do well to spend some time reading, developing, and understanding key ideas. Byers asked the panelists how they would apply this strategy, given their experience and relation with NASA.

Klug-Boonstra explained that she and her colleagues have had to reassess the issue of evaluations. They conduct deep evaluations with their conferences and workshops to understand what will be helpful, and they have learned that incremental and small things are paying off. They make an attempt to understand who the audience will be in terms of teachers, the grade groups, the level, and local knowledge. They also ensure that they reconnect with teachers. The teachers become the students and go through what their students will experience. She explained that having people who have been in the classroom help to design the professional development results in trust from teachers.

Milano explained that Orange County Public Schools focuses a lot on developing evidence statements with teachers because the district in which she works is used as a basis for proficiency determination. She added that using rubrics against which student progress is judged helps in understanding where students are going; thus, this is an effective mechanism. She then explained that one of the things that her district has done is keep a core group of teacher leaders employed on assignment at the district office for partners to work with directly. She stated that the method of going from school to school can have a small-term impact but not a large-term impact. She stated that she wished that more partnerships included the idea of analyzing assessment tasks.

DeCharon explained that shifting to the thematic approaches, as opposed to mission-specific approaches, will make NASA more flexible in serving the professional development needs of educators. She stated that the science and engineering practices used in the classroom could be modeled after the authentic work being done by NASA professionals. She stated that many scientists and engineers are beginning to realize that they need to learn about how the classroom operates and how teachers do their job in order to work with them effectively as partners.

Penuel stated that part of working with districts is understanding that there is a need to work by *persuasion*, and they rarely have the ability to command that teachers attend professional development. "But pre-coaching your speakers, pre-preparing your speakers in terms of the languages that the teachers are using, in terms of the kinds of professional development they need," is vital, Penuel said. "And making sure that the scientists can also adopt that language and then use their relevant, real-world examples to tie right into the concepts that you're teaching" is valuable, he added. "It is not just science talk and science lingo—but actually connecting with the way teachers need to meet those standards or those crosscutting concepts or those practices."

Klug-Boonstra mentioned the evolution of understanding the needs of teachers, and an opportunity associated with this is bringing in subject-matter experts. She stated that NASA has a great opportunity to make that connection with teachers.

Byers asked the panelists' opinions regarding co-design for curriculum with teachers. DeCharon began with a discussion of concept maps. She explained that her group has trained more than 250 scientists and engineers in concept mapping, which consists of deconstructing complex science into discreet concepts and describing their relationships. She explained that she and her team have worked with scientists and engineers and have had them break down their science into concept maps, followed by workshops and webinars where they sit down with teachers or educators. She stated that scientists have a way of presenting their information that often is not well-matched with what students need—the different cultures problem that was raised by many panelists throughout the workshop. Thus, she said, teachers can help inform scientists, which results in the scientists being able to see their work in the context of the needs of society in classrooms. She noted that many scientists have redone their own teaching and that they have re-organized their entire curriculum for their classrooms.

Milano explained that her group is involved in curriculum design or co-design every summer. She explained that many times people come in and redesign a curriculum for a week or two over the summer, which has value. However, because it only touches one course, the district is often left without the ability to replicate that in multiple courses. She provided the example of the "Bridge to STEM" program, which was an early-childhood, pre-physics program that took research from the National Association for Education of Young Children and the National Science Foundation and focused on how to teach children about the force of motion in pre-kindergarten and on through first grade. A small group of teacher leaders from the early childhood community worked closely with university professors and science, technology, engineering, and mathematics (STEM) professionals to design lessons, test them in classrooms, and go through the revision process based on what was learned from students, teachers, and parents, which resulted in a firm curriculum. She then discussed a positive example of co-design, which was a project called Moving STEM into the Main STREAMS, whose mission was to determine how to get STEM learning into all subject areas. NASA and Kennedy Space Center hosted virtual field trips and face-to-face experiences in which people were brought out and shown what types of jobs students can obtain directly out of high school or college. Feedback was provided regarding what skills were missing, and lessons were developed based on that gap analysis.

An audience member made a comment about professional development and said that the scientist/educator partnership benefits both sides. He asked Klug-Boonstra if she is able to bring scientists along for professional workshops. Klug-Boonstra responded that her group brings along graduate and undergraduate students to be a part of professional development to expose them to and have them benefit from the kind of learning that is going on.

Penuel provided detail about the curricular co-design work in Denver. He stated that his group is gaining access to "break down the walls" of schools so that kids can do science and connect to the community, which is the heart of the vision behind citizen science.

Another audience member asked the panelists about recommendations or thoughts on the role of NASA with regard to the challenges, opportunities, and needs of pre-service teachers. Penuel stated that his group provides its pre-service teachers with concrete images of what three-dimensional science learning looks like. He explained that curriculum analysis is also important. His group prepares teachers with critical analytic skills. He posed the question of how to form a network of support for a new teacher to learn who are the experts and great science teachers in their particular building or district. He stated that support is needed due to a disconnect between the experience of most pre-service teachers and the expectations placed on them when they go into a district.

An audience member mentioned that he used to go out into community high schools and middle schools to discuss science and encourage students to go to college. He mentioned that the late Sally Ride had emphasized that the problem is mathematics because this is what will cause students to drop out, not the science. He specifically mentioned that this is a larger problem for underrepresented minorities and girls and young women.

Penuel replied that it is fundamental to engage children in computational thinking about systems. He said the crosscutting concept of scale has a deeply mathematical focus and it is relevant to NASA. He explained that educators are losing math students, partly at the undergraduate level, because children in science and engineering are not substantively involved in solving complex problems. Milano stated that her group struggles with computational thinking, and it is constantly looking for more authentic ways to bring in real-world experiences.

An audience member brought up the topic of long-term partnerships. She stated that in NASA Science Mission Directorate public outreach, they have enjoyed long-term stability. However, due to the potential funding model change, this stability may be lost, and the ability to do long-term partnering may be lost as well. She asked Milano, as well as the other panelists, what can be done if this ability is lost. Milano responded by stating that there is meaningful work that can be done if it is not possible to engage in a long-term partnership. She provided an example of engaging in the development of virtual field trip models. She stated that in her position, one of her roles is to curate digital learning objects. She suggested developing professional development programs around how to use digital learning objects from NASA, assisting schools districts with reviewing their own curriculum, and moving training to a blended format or using online courses because the face-to-face delivery model is often not cost-effective.

Penuel explained that one component of going after competitive grants is figuring out what new problem can be addressed through a partnership. He stated that there are new funding programs for research practice partnerships. He often discusses the significance of entering a partnership and figuring out what problems will be worked on that will require renegotiation in school districts.

An audience member stated that there was a lack of discussion at the workshop so far about elementary-level professional development. He stated that there are many science-phobic and math-phobic elementary teachers. Penuel responded by referencing work being done by his colleagues at the University of Washington focused on adapting kit-based science. His colleagues are doing professional development work focused on adapting kits, for the NGSS world, to incorporate scientific explanation and engagement in argumentation. He is also seeing focused professional development at the elementary level all over the country. Klug-Boonstra explained that science-phobic teachers are afraid that they will be asked a question they cannot answer. Thus, in professional development they create a flipped classroom in which kids can be enabled and teachers are free from the fear of not knowing.

An audience member commented that NASA brings more than just wonder. NASA builds things that have never been built before. The audience member then transitioned into professional development and suggested that it may be useful for NASA to engage in professional development for teachers that is focused on mission events such as the New Horizons mission's flyby of Pluto and the Juno mission's fly-by of Jupiter (Figure 5.1). She asked the panelists to discuss short-term versus long-term professional development. DeCharon explained that a solution could be looking at all of education and public outreach as a team and thinking of missions as collective, as

FIGURE 5.1 New Horizons' flyby of Pluto in 2015 is an example of an event that can be used in science classrooms. SOURCE: Courtesy of NASA/JHU-APL/SwRI/Steve Gribben.

opposed to each individual mission event at a specific moment. However, she emphasized taking advantage of a moment to get people's interest. Klug-Boonstra explained that her group's approach was to have children become "eReporters" and have them find events that are going on in the world. Thus, the children are required to find the stories and determine which ones are important to bring recent science events to the attention of the class.

Penuel brought up the issue of equity with events-based work. He explained that when teaching is driven by events, schools under the least amount of pressure with the most advantaged students are able to do events-based work while other schools cannot.

6

Evaluation of Education and Evaluation in Practice within NASA SMD

Part 1 of Session 5 focused on theory concerning the guidelines, and Part 2 focused on evaluation and practice.

PART 1—EVALUATION OF EDUCATION

Moderator	Theresa Schwerin, Institute for Global Environmental Strategies
Speakers	Martin Orland, WestEd
	Steve Schneider, WestEd

Making the Right Choices: How to Get the Most Value out of eVALUation!

Martin Orland and Steve Schneider, WestEd

The presentation prepared by Martin Orland and Steve Schneider, both of WestEd, was delivered by Orland. He began by stating that at its core, scientific inquiry is the same in all fields. It is "a continual process of rigorous reasoning, supported by a dynamic interplay among methods, theories, and findings. It builds understandings in the forms of models or theories that can be tested," he said. Orland commented that during the workshop, participants have spoken a lot about evidence and how to encourage a passion for evidence among students. He suggested that an equal agenda and passion are needed in terms of the nature of scientific inquiry and education.

Orland explained that the Common Guidelines for Education Research and Development (generally referred to as "the Guidelines") that were jointly developed by the National Science Foundation (NSF) and the Institute of Education Sciences of the Department of Education are an attempt to demystify scientific processes as they are applied to educational research and evaluation.[1] The Guidelines are a cross-agency framework that describes broad types of research and development (R&D) and "the expected purposes, justifications, and contributions of various types of agency-supported research to knowledge generation about interventions and strategies for improving learning."

Orland said the Guidelines are necessary because the American education system needs research to produce stronger evidence at a faster pace. Constrained federal resources demand that NSF and the Department of Educa-

[1] Institute of Education Sciences, U.S. Department of Education and the National Science Foundation, "Common Guidelines for Education Research and Development," 2013, pp. 1-53.

tion and other agencies purposefully build on each other's research and development portfolios. The Guidelines provide a cross-agency vocabulary and set of research expectations that are critical for effective communication.

Knowledge development in education is not strictly linear, Orland said. There are three categories of educational research: core knowledge building, design and development, and studies of impact. These all overlap. This requires researchers and practitioners representing a range of disciplines and expertise. It may require more studies for basic exploration and design than for testing the effectiveness of a fully developed intervention or strategy. It also requires assessment of implementation and not just the estimation of impacts. Finally, it includes attention to learning in multiple settings, both formal and informal, Orland explained.

The Guidelines are organized according to the following:

- *Purpose.* How does this type of research contribute to the evidence base?
- *Justification.* How should policy and practical significance be demonstrated? What types of theoretical and/or empirical arguments should be made for continuing this study?
- *Outcomes.* Generally speaking, what types of outcomes (theory and empirical evidence) should the product produce?
- *Research plan.* What are the key features of a research design for this type of study?
- *External feedback plan.* A series of external, critical reviews of project design and activities. Review activities may entail peer review of the proposed project, external review panels or advisory boards, a third party evaluator, or peer review of publications. External review should be sufficiently independent and rigorous to influence and improve quality.

Orland said that the Guidelines will not preclude innovative projects. They are intended to help principal investigators in proposal preparation. He said the key point is to ensure that projects are explicit about their research questions, methods, and analytic approaches in their proposals. The criteria should be relevant for all types of education R&D efforts. The Guidelines can help practitioners develop a better understanding of what different types of education research should address and might be expected to produce, he said.

Orland explained that the Guidelines apply to proposals, but they foreshadow what will come from the R&D effort. Each section of the Guidelines is connected to evidence of some aspect of the proposal and the proposed work. Throughout, the Guidelines provide explicit and implicit messages about what counts as evidence and what needs to be considered.

Orland concluded by noting, "So in a sense, you are building the airplane while you are flying it. You are both doing this exploratory research, and you are learning as you are going, so that you are refining and improving."

PART 2—EVALUATION IN PRACTICE WITHIN NASA SMD

Moderator Theresa Schwerin, Institute for Global Environmental Strategies
Speaker Hilarie Davis, TLC, Inc.
Panelists Bonnie Eisenhamer, Space Telescope Science Institute
Jenny Gutbezahl, Brandeis University
Frances Lawrenz, University of Minnesota

Using Evaluation to Increase and Measure the Impact of Education

Hilarie Davis, TLC, Inc.

Hilarie Davis of TLC, Inc., started Part 2 of Session 5 by showing an objective from the NASA's 2014 strategic plan:[2]

[2] NASA, *NASA Strategic Plan 2014*, Washington, D.C., p. 34.

> Advance the nation's STEM [science, technology, engineering, and mathematics] education and workforce pipeline by working collaboratively with other agencies to engage students, teachers, and faculty in NASA's missions and unique assets.

Davis discussed the barriers to evaluation identified from surveys of NASA education specialists. These include the following:

- Evaluation seems like it is being done for someone else [outside of the school system] instead of for improving the program;
- The evaluation topic is not close enough to the work being done to be meaningful to the teacher;
- The evaluation is not realistic in its scope or methods;
- Evaluations are too costly for the perceived value; and
- The evaluations feel like an audit or judgment of the people and/or program.

She said that evidence indicated that the barriers could be overcome. NASA's Science Mission Directorate (SMD) held forums at which more than 200 people attended evaluation sessions. The attendees reported that the session had significantly affected their understanding of evaluation, their perception of its value, and their intention to use it in the future. They identified strategies for overcoming barriers. These included the following:

- Embed evaluation in the whole project cycle—provide feedback and support for this;
- Give the evaluation credibility by involving the stakeholders appropriately;
- Build the evaluation around questions that are important;
- Use reasonable, practical approaches to collect data;
- Be clear about the purpose of the evaluation; and
- Use the results of the evaluation to guide decision-making about program elements, goals, and funding.

An example was given of a valuable evaluation that was done for seventh and eighth graders using results

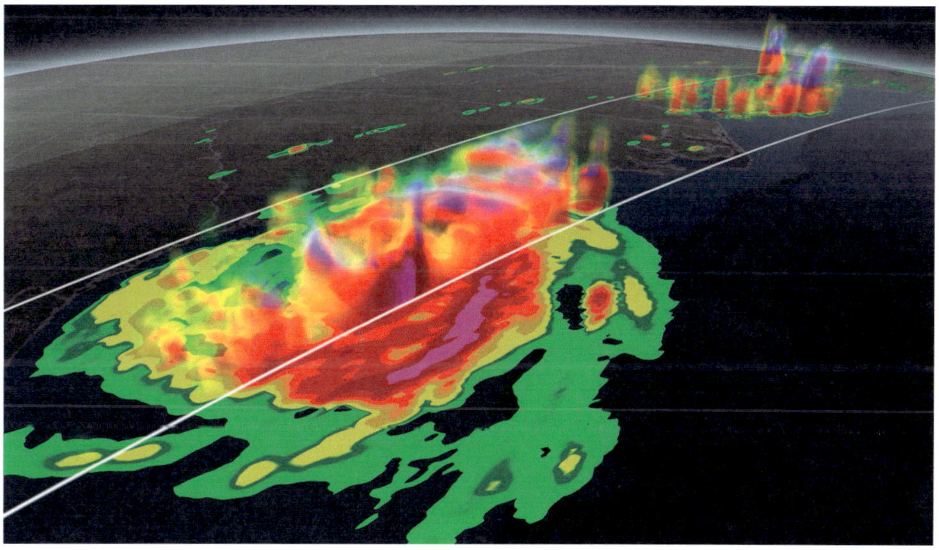

FIGURE 6.1 An example of data produced by NASA's Global Precipitation Mission (GPM). GPM was used in curriculum taught to students as part of a test on how well they retained concepts over time compared to the standard curriculum. SOURCE: Courtesy of NASA's Scientific Visualization Studio; data provided by the joint NASA/JAXA GPM mission.

from the Global Precipitation Measurement mission (GPM) (Figure 6.1). Curriculum concepts were taught to one group of students using GPM as an example. Then the students given the GPM example were compared to general curriculum students. The two groups did equally well on knowledge tests for seven lessons during the year, but the students given the GPM example did *better* than the general curriculum students on the retention of concepts in an end-of-the-year test. The retention of the science curriculum was enhanced by the context.

Davis mentioned the intern program at NASA Ames Research Center, where rural high school students worked with Ames astrobiologists studying extremophiles in nearby Lassen Volcanic National Park. Before they started, and at additional times during the year, the students answered core questions about astrobiology that the science team used to guide their interactions with the students. The students collected and analyzed data. The students later presented their findings to the community to demonstrate their understanding of the science.

Davis also discussed several other NASA projects, including the Magnetosphere Multiscale Mission, the Heliophysics Education Ambassadors, and the Adler IBEX After School Club.

Davis said that evaluation can be as important as the work itself, and whenever possible one should include an evaluator at the beginning stages of designing new projects. "There is no point in doing the work unless you can prove its worth," she noted.

As to why the strategies employed to overcome barriers work, she listed several reasons:

- People like feedback—not judgment. Judgment feels punitive while feedback feels helpful.
- People want to do well—they set out to succeed, not to fail—so they appreciate a fair assessment that may help them improve.
- Evaluation throughout the project cycle improves it every step of the way, so there are a lot of chances to improve.
- People want answers to their questions, so when they help develop the questions, they care about the answers.
- People improve when they have a clear path to getting better, which is why they say the project cycle rubric helps.
- People delivering programs know where and how good data can be most effectively collected.
- Evaluators do a better job when stakeholders evaluate their evaluation plans, methods, and measures for value and validity. Stakeholders are also experts.
- Decisions based on good data about a program are honest and productive; decisions made without good evaluation data are suspect and feel arbitrary, which discourages productivity.

Davis concluded by stating that "through evaluation we are able to collect evidence and develop explanatory models of how to bring back the wonder for teachers and students to know, care about, and pursue NASA and STEM learning."

PANEL DISCUSSION

Hilarie Davis and Theresa Schwerin joined Bonnie Eisenhamer, Jenny Gutbezahl, and Frances Lawrenz for the panel discussion. The organizing committee developed the following guiding questions to provide focus to the panel discussion:

- Why and how does NASA evaluate the programs it executes?
- What are examples of evidence that the evaluation of NASA's programs is providing useful information to improve the programs?
- How does NASA make a difference in STEM education, and how is this known?
- What are the greatest challenges or barriers that people have encountered related to SMD education evaluation? What strategies have been used or recommended for addressing these barriers?
- How does the evaluation of NASA programs compare to the model presented for education by Orland and Schneider in Part 1 of this session?

- What is the mechanism by which the results of evaluation change NASA education programs?

Lawrenz started the session by noting that that evaluation has been underfunded. There have also been unrealistic expectations and an inability to address "real" questions as well as sampling bias. But new models are available for NASA. He suggested that "less is more" and that it is important to be selective about which battles to fight.

Eisenhamer said that it is important to "plan evaluation with your end goal in sight." In addition, it is important that

- Evaluation questions are well matched to define purpose and strategies; and
- Evaluation questions and methods are appropriate to the stage and maturity of the program.

Eisenhamer noted that front-end planning is needed, and evaluation is planned with the end outcomes in mind. A new direction for SMD education will require the involvement of an evaluator from the beginning of the program, she said.

Gutbezahl was involved in the 1997-2000 NASA Space Science Education and Public Outreach effort and is seeing now many of the same challenges she observed more than 15 years ago. These include a culture clash between scientists and educators, a lack of coordination across the system leading to gaps and redundancies, and challenges between going for depth versus breadth. She said that she is also seeing many of the same strategies to address these problems, including creating common goals to overcome culture differences and going to the users to discover their needs.

Gutbezahl also noted that current evaluation places more emphasis on "empirical evidence," meaning numbers "which are really not any more empirical than qualitative data," she said. This leads to an "emphasis on breadth, because it is easier to count noses than measure true impact."

There are often two outcomes: not enough data on what works, or great data showing that something does not work. The panelists stressed that the issues are long-term and have to be addressed as such.

7

Enabling Actions

The Session 6 moderator was education consultant James Manning. Manning explained that the workshop participants were to split into five breakout groups to discuss the previous sessions and identify issues and relevant areas. Each of the breakout groups had guiding questions that they were to consider. The groups reconvened after 1 hour to share their views.

ALIGNING TO STANDARDS

The Breakout Group 1 speakers were Brett Moulding of the Utah Partnership for Effective Science Teaching and Learning and Heidi Hammel of the Association of Universities for Research in Astronomy. The group's guiding questions were the following:

- What actions can NASA take to build upon, leverage, and/or expand its current efforts to align to and support the new vision described in the National Research Council's *A Framework for K-12 Science Education*[1] (generally referred to as "the Framework"), the Next Generation Science Standards (NGSS), and other standards initiatives?[2]
- What new opportunities can be explored, and what challenges need to be overcome?

Hammel began by stating that the group came up with several key points. She emphasized that the overall summary of the group is the focus on the NGSS and the interaction with NASA Science Mission Directorate (SMD). Among the issues the group discussed were the following:

- The NGSS and the Framework support consistency across states, which will be helpful for NASA's broader program.
- Combining science with other subjects (e.g., math, literacy, engineering, and art) will result in better alignment to the standards of the Framework.

[1] National Research Council, *A Framework for K-12 Science Education: Practices, Crosscutting Concepts, and Core Ideas*, The National Academies Press, Washington, D.C., 2012.

[2] See the "Introduction and Background" chapter for a discussion on the Next Generation Science Standards and the Framework for K-12 Science Education.

- Linking to everyday phenomena can help students if they can understand how science is important in their lives.
- NGSS lesson templates aid in creating lessons that are well crafted.
- There is a need for multiple assessments throughout the process.
- There is a need to bundle or organize multiple resources across disciplines and programs that would help in developing broader-based curricula.
- It is important for NASA SMD to focus on aligning professional development material and curriculum resources to the NGSS, including collaboration with evaluators.
- There is a need for NASA SMD materials development specialists, which include scientists, engineers, education and public outreach specialists, and evaluators, to be informed about NGSS and the Framework.
- There is a need for performance expectations to simultaneously match the math levels with the science levels.

On the last point, the breakout group discussed at length the nature of mathematics and science instruction and the need to match it to the NGSS performance expectations. Hammel explained that the group concluded that alignment is done best with a team.

An audience member asked a question about whether NASA ought to produce examples of best-practices lesson templates or if they [the group] were saying that NASA education and public outreach need these templates in order to move ahead. Brett Moulding responded by stating that there is a need to develop a template so that lesson delivery is consistent to ensure that teachers do not have to re-interpret a different format. He also emphasized that who develops the template is not as significant as its wide and consistent use.

Hammel stated that she would present slides pertaining to a summary of the presentation delivered by Stephen Pruitt and the panel discussions. She broke the presentation into two components: What are the NGSS? and What are the NGSS implications for NASA? Hammel described the former as a vision in which students are at the center of performances that demonstrate their understanding of doing science. She emphasized that the goal is that all students learn and enjoy using science to make sense of phenomena.

Hammel then addressed the implications of the Framework and the NGSS for NASA. She explained that NASA will need to change some of its materials to ensure that programs and lessons are more like the Framework and the NGSS. She noted that the National Science Teachers Association EQuIP rubric would help in accomplishing this.

Hammel moved on to the panel discussions. She began with the Session 1 panel and noted that the role of NASA is to answer the questions of how and why. NASA provides real-life resources and direct examples of what science, technology, engineering, and mathematics (STEM) career pathways can be. She noted that Maya Garcia of the Office of the State Superintendent of Education for the Government of the District of Columbia did a good job providing a teacher's voice on how one can transition the NGSS into the Washington, D.C., school district. Garcia addressed the question of how the Framework can reach all students. The answer to this question is the Framework's flexibility. Hammel also noted that Holly Ryer of the Space Science Telescope Institute discussed how the NGSS focuses on real-world activity; and Sam Shaw of the South Dakota Department of Education used South Dakota as an example of a state that will use the Framework to develop its own standards.

CURRICULUM SUPPORT RESOURCES

The Breakout Group 2 speaker was Richard McCray of the University of California, Berkeley. The group's guiding questions were the following:

- What actions can NASA take to build upon, leverage, and/or expand its current efforts to translate its science into curriculum-support materials and resources for formal and informal education and to encourage educator use?
- What new opportunities can be explored, and what challenges need to be overcome?

McCray began by explaining that NASA has great curriculum development and teacher training examples; however, many of the examples do not scale up to the entire nation. He provided an example of a poster focused on allowing teachers, each with several high school students, to study infrared astronomy for a full year, including

2 weeks at NASA Ames Research Center, participation in a flight on SOFIA (Stratospheric Observatory for Infrared Astronomy), and presentation of results at a meeting of the American Astronomical Society; it is a wonderful program that does not scale up due to the cost of $50,000 per teacher.

He then posed the question of how to get NASA resources to teachers. He stated that the solution is better partnerships with textbook publishers and with state curriculum developers. He emphasized that if NASA is developing teacher resources, then it needs a plan for how to make partnerships. He also noted that the National Science Foundation (NSF) does a good job with encouraging its education developers to work with commercial entities, which is something that NASA can learn from NSF.

McCray then addressed the topic of information technology access among different socioeconomic groups. He mentioned that more than 95 percent of upper socioeconomic groups have iPhones or tablets, whereas only 65 percent of the poorest individuals have access to these technological resources.

He also discussed information technology as it relates to teacher training. A workshop participant mentioned that massive open online courses can be an effective tool in teacher preparation for the use of new material.

An audience member commented that his group recently finished a project in partnership with a publisher to develop cosmology curriculum for freshmen-level and college non-majors, and he believes that this information will reach a lot of people. He emphasized a point made earlier by McCray that there is useful material that is not being disseminated well.

McCray then discussed ongoing changes in the publishing industry and that the workshop community would have to examine whether copyright and intellectual property ownership presents an obstacle to the partnerships that it needs.

COLLABORATIONS

The Breakout Group 3 speaker was Greg Schultz of the Astronomical Society of the Pacific. The group's guiding questions were the following:

- What actions can NASA take to build upon, leverage, and/or expand its current collaborations among scientists, teachers, and formal and informal education institutions?
- What new opportunities can be explored, and what challenges need to be overcome?

Schulz began by reporting that the group talked a lot about collaborations and partnerships. He referenced a panel on the first day of the workshop in which a participant stated that the Space Telescope Science Institute has documented approximately 500 currently ongoing partnerships between the institute and other groups and institutions. (See Figure 7.1.) He said that some are large organizations and others are with a school district or local efforts. Thus, categorizing partnerships and what they really entail is vital to understanding them.

Schulz then stated that the group discussed individual activities or partnerships that people do on their own time. He highlighted individuals that work at NASA centers who allocate time to education and public outreach, and he emphasized that there could be more encouragement to do that. He also stated that the group thought that scientists would benefit from training that explains educators' needs and the context for the NGSS. The encouragement could focus on starting small and growing activities. He explained that one of the group members mentioned that scientists should be reminded of opportunities to interact with the public, even if the opportunities are not funded through NASA.

Schulz explained that the group talked about making connections with other parts of NASA, such as the Office of Education or the Office of Communications, to better inform activities and avoid duplicating efforts. He stated that another advantage of forum-like organizations could be fostering larger partnerships than those brought together by individual scientists or smaller projects.

He also added that the NASA education forums bring together scientists and educators. The forums can also bring together those in the education workshop community for professional development opportunities, dissemination of findings, and sharing of best practices. Through this, they can learn to align their products and programs with the NGSS, Schulz said.

FIGURE 7.1 Two Hubble Space Telescope images produced in 1995 and 2014. The Space Telescope Science Institute has documented many partnerships with small and large organizations such as the partners involved with the creation of the Eagle Nebula images. SOURCE: 1995 image courtesy of NASA, ESA, STScI, and Paul Scowen and Jeffrey Hester of Arizona State University; 2014 image courtesy of NASA, ESA, and the Hubble Heritage Team (STScI/AURA).

An audience member encouraged those in the audience to think about ways in which the broader science community can be engaged in education and public outreach activities. Schulz agreed and stated that a culture shift needs to be made. He explained that there has been more willingness within the past 20 years among younger scientists who want to do better education and public outreach. If organizations can encourage their scientists to invest their time in this, then everyone can benefit.

PROFESSIONAL DEVELOPMENT

The Breakout Group 4 speaker was Albert Byers of the National Science Teachers Association. The group's guiding questions were the following:

- What actions can NASA take to build upon, leverage, and/or expand its current efforts to provide professional development support to pre-service and in-service teachers and informal educators?

- What new opportunities can be explored, and what challenges need to be overcome?

Byers began with the group's discussion of NASA actions for professional development efforts. He stated that one way to improve professional development is to define best or promising practices or program development models from research. He explained that the group also wanted to communicate that there is still value in small programs. The group also discussed the value of inventorying the best models in order to disseminate them across the agency electronically. Byers explained that the inventory then needs to be disseminated to school districts. He stated that the group also suggested that there was value in focusing on a common needs assessment with research-based evaluation models. The group discussed using the K-12 Science Education Framework as the foundation for program development.

On the issue of new opportunities and new challenges, Byers stated that one of the big challenges the group identified was scientific literacy for underserved populations and adults. He explained that the group discussed interagency collaboration, which ties into the CoSTEM (Committee on Science, Technology, Engineering, and Math Education) effort with respect to professional development. They also talked about intra-agency (within NASA) collaboration and across missions using the Framework.

An audience member noted that professional development that includes actual scientific research is intensive and expensive. Another audience member highlighted that Byers' group discussion did not include professional development for scientists about education. She stated that bringing educators to conferences to talk about the standards would likely be well received. Another audience member then explained that one of the opportunities in program development is the potential for collaborations with the Office of Education, which is something that the education community needs to do.

EVALUATION

The Breakout Group 5 speaker was Theresa Schwerin of the Institute for Global Environmental Strategies. The group's guiding questions were the following:

- What actions can NASA take to build upon, leverage, and/or expand its current efforts in measuring and assessing its impact in science and engineering education?
- What new opportunities can be explored, and what challenges need to be overcome?

Schwerin began with a list of the group's responses to the first question regarding NASA actions to build upon its impact in science and engineering education. Their first solution focused on being clear about evaluation requirements from the start, when a project is funded. The group also suggested standards for evaluation across a program to ensure that there is a clear understanding of what is expected when it is proposed. She then explained that the group also thought it was important to measure implementation and the end result. It also suggested investigating the feasibility of providing evaluators that the community can access.

Schwerin then addressed the second question on new opportunities and challenges. She first discussed the Wavelength system[3] and the need to examine how it is being used, as well as the assets contained within that system. She stated that creating a registry that includes teachers and students and a way for them to create profiles to apply for programs is a way to track individuals and obtain good, rich data. She provided the example of the Office of Education's one-stop shop for undergraduate and graduate fellowships. She explained that some of the anecdotes at the workshop highlighted its success. Students like it because they have one place to go to apply for programs. She mentioned the possibility of creating this kind of registry for teachers to use.

The group also suggested pulling together findings across projects to have a collection of results and evaluation methods. Schwerin mentioned that the group talked about communication and how the dialog established at the forum meetings can be continued. It discussed the solution of having annual reviews of education projects and ensuring that managers and policy makers have the information they need. The group recommended using

[3] See the "Introduction and Background" chapter for a discussion on NASA Wavelength.

current information technology to collect, report, and make data available. Schwerin posed the question of whether the workshop community can collect not only data that stakeholders and Congress want to know, but also deeper evaluation data.

The group also discussed bringing in representatives from projects and programs to report results to policy makers and administration. Schwerin explained the significance of finding a way to let the people doing the projects tell their story. The group also suggested having contractors collect and analyze data. Schwerin then moved on to the creation of a repository of questions and measures (i.e., evaluation instruments) that have been approved.

Schwerin stated that the group talked about taking the lessons learned, which was a theme across the workshop. She noted the significance of having one index of resources and pushing information out to many different places. She suggested building the NASA SMD education and public outreach community out further.

An audience member commented that previously there was talk about the difficulty in gaining access to student data that are protected; however, many larger districts have research departments and institutional review boards. He explained that his group has worked with districts before, and sometimes the data may be obtainable. He stated that if there is collaboration with districts, then the information is available without requiring students to self-register.

Another audience member suggested that it would also be helpful to have people who are experienced in education or learning research available as advisors so that information can be drawn and research can be put into practice.

Finally, an audience member commented that there does not seem to be an attempt to provide an overall story of how education is effective for NASA. He believes that this is the critical piece around which to focus a strategy. He noted that SMD is one piece of the conversation happening across Congress, the Office of Management and Budget, and the nation.

8

Wrap-Up

Brett Moulding, co-chair of the organizing committee, began the summary and wrap-up by thanking all of the workshop participants. Moulding explained that he and Phil Christensen, the other co-chair, would provide a personal recap of each workshop session and discuss several points that were made within each session.

The first session focused on the vision for science education. Moulding explained that whenever members of the education community say "Framework," they mean the *Framework for K-12 Science Education*.[1] Moulding added that all states will align to the Framework according to their own cycles.

Christensen then explained that they would present five slides that did not exactly follow the five sessions, because their goal was to provide a higher-level review of the workshop. He emphasized that their summary was not meant to summarize the sessions or the workshop; instead, it was a summary of Christensen's and Moulding's thoughts and observations.

Christensen noted that there was a lot of discussion about curriculum development, and an overarching theme was that the NASA Science Mission Directorate could benefit by organizing curriculum materials around the Framework. He also emphasized the importance of moving away from basic content elements (e.g., activities, lessons) into broader pieces of curriculum that will aid accessibility for teachers. Christensen stated that as communicators, they [the workshop community] could work with their education partners to help organize material that will facilitate greater conceptual learning. He also made the point that the science community has focused on mission-specific or instrument-specific materials for a long time. He suggested moving away from mission-specific materials to standards-based, crosscutting concepts and disciplinary core ideas so that, for example, people working with the Hubble Space Telescope can incorporate data from Cassini at Saturn or a Mars mission. By helping to develop materials that teachers can use in a three-dimensional world and encouraging students to think more about science and engineering practices, space and Earth scientists can contribute to the next generation of curriculum.

Moulding then stated that over the course of the workshop, participants talked about the ability of partnerships to address the issue of scalability, and he proposed that the audience think about projects in terms of a system that happens to be one of the crosscutting concepts. Thus, the idea of scalability relates to the interface between the teacher and the student. Those in the audience are the "support staff" for the interface where teaching and learning happens between a teacher and a student. He told the audience members to recognize that effective partner-

[1] National Research Council, *A Framework for K-12 Science Education: Practices, Crosscutting Concepts, and Core Ideas*, The National Academies Press, Washington, D.C., 2012.

ships take time to fully develop, and they must be built on equitable and mutual needs. He urged entering into the partnership relationship with an open mind and thinking about how the partnership will benefit local children as well as local education goals. Moulding emphasized that key elements of an effective partnership are the education professionals—the boundary-spanners. The translators from science to education have an essential role that is very much needed. He also noted that existing partnerships must inform each other of changes in education or in science. He stated that just because a partnership today is effective does not mean that it will be effective in the future.

Christensen mentioned the workshop discussion about changes in funding, as well as the new cooperative agreement notice. He encouraged his colleagues to embrace the changes coming to science education. He stated that NASA could play a key role in professional development for teachers. However, this professional development needs to meet the standards established for science education within the Framework. Christensen added that it would be beneficial for those on the science side to learn more about the Framework. He also emphasized the ongoing and iterative nature of professional development, and noted that things will continue to evolve. He stated that the audience members should be informed by student learning, be a part of local organizations, and allow for teacher autonomy. Furthermore, the community must constantly remember that teachers are individuals with their own styles.

Moulding referred to the National Association of Research and Science Teaching white paper on professional development standards.[2] He said that the Council of State Science Supervisors is revising standards to align to the Framework.

On the subject of research and evaluation, Moulding explained that guidelines are changing the evaluation expectations for educational programs. He emphasized that evaluation is most powerful when it is program specific and also aligned to evaluation standards. Moulding said that historically, evaluation is underfunded. Moulding stated that evaluation is *not* research; thus, the models used in professional development have to be based on current educational research. The community is hungry for professional development, instructional material, and classroom instructional strategies research that are effective at improving student learning. Moulding concluded by stating that when data and results are shared, the entire community of scientists and educators benefit.

Marc Allen of NASA headquarters then spoke briefly and encouraged audience members to send their questions or comments regarding the cooperative agreement notice to the official [NASA] e-mail address. Michael Moloney, director of the National Research Council's Space Studies Board, concluded by thanking the workshop organizing committee for helping the staff put together the agenda. (See Figure 8.1.)

---

[2] E.R. Banilower, J. Gess-Newsome, D. Tippins, "Supporting the Implementation of NGSS through Research: Professional Development," National Association for Research in Science Teaching, Reston, Va., http://www.narst.org/NGSSpapers/Professional_Development_%20June2014.pdf.

FIGURE 8.1 The 2-day workshop held in the auditorium of the National Academy of Sciences building featured extensive discussion and participation by a broad range of experts in the science education field. SOURCE: Harrison Dreves, NRC.

Appendixes

A

Statement of Task

An ad hoc committee will plan and conduct a workshop to discuss maximizing the effectiveness of the transfer of knowledge from the scientists supported by NASA's Science Mission Directorate (SMD) to K-12 students directly and to teachers and informal educators. In addition, the connection between these SMD efforts and the implementation of the National Science and Technology Council's Committee on Science, Technology, Engineering, and Mathematics (CoSTEM) STEM Education Strategic Plan will be discussed.

The workshop will focus not only on the effectiveness of recent models for transferring science content and scientific practices to students, but will also serve as a venue for dialogue between education specialists, education staff from NASA and other agencies, space scientists and engineers, and science content generators. Workshop participants will review case studies of scientists or engineers who were able to successfully translate their research results and research experiences into formal and informal student science learning, including measureable improvements to student achievement and other valued outcomes listed below. These may include a variety of outcomes, from awareness to conceptual understanding, from understanding discovery to the nature of the scientific enterprise, and from the joy of learning to authentic involvement with actual scientists. Education specialists (e.g., state science supervisors, teacher education departments, professional development providers, informal science institutions/organizations, such as planetariums, after-school STEM education providers, state STEM education networks, and education researchers) will share how the science can be translated to education materials and directly to students, and teachers will share their experiences of space science in their classrooms.

Amongst the things the committee will consider in assembling the workshop are issues such as:

- Where is the intersection between what NASA space science can provide for education (as part of the CoSTEM STEM Education Strategic Plan) and what education providers want and need from the NASA science community? What can NASA do to better understand educators' and education providers' needs?
- How can scientists, engineers, and education providers work together to improve the science education of school-aged students? What are the common goals? How are the goals of educators and NASA scientists different? How can these differences be bridged?
- How is it determined if a program has been successful? Do scientists and education experts agree on what constitutes success? How well does SMD's process of development, trial, evaluation, and performance measurement support an "evidence-based" approach?

- What are institutional arrangements that provide effective platforms for successful cooperation or collaborations? What are the barriers, if any, to meeting the common goals?
- What are existing examples of successful NASA SMD education efforts?
- How can NASA work with/lead other agencies to have its successful education efforts reach a wider audience? How can this work be targeted towards students who may need it most?
- How should success of NASA education efforts be defined and measured? How can NASA determine whether its education efforts are having a measurable long-term effect on student achievement and involvement in science?

The workshop will feature invited presentations and discussions that encourage broad audience participation. The committee will select and invite speakers and discussants and moderate the discussions. An individually authored summary of the presentations and discussions at the workshop will be prepared by a designated rapporteur in accordance with institutional guidelines.

B

Workshop Agenda

DECEMBER 2, 2014

8:30 a.m.	**Welcome and Introduction** ***Outlining the Goals of the Workshop***	Michael Moloney, Space Studies Board Director *Committee Co-Chairs:* Phil Christensen, Arizona State University Brett Moulding, Utah Partnership for Effective Science Teaching and Learning
8:40	***Welcome from NASA***	Kristen Erickson, NASA Science Mission Directorate
8:50	**Introduction to the Keynote Speaker**	
8:55	**Keynote Presentation:** ***Sharing the Adventure with the Student: How Do Authentic Experiences Reach Students***	John Mather, NASA Goddard Space Flight Center
	Interaction with the Audience	
9:50	**Setting the Stage**	*Panelists:* NASA Education Forum Leads Laura Peticolas, University of California, Berkeley Theresa Schwerin, Institute for Global Environmental Strategies Stephanie Shipp, Lunar and Planetary Institute Denise Smith, Space Science Telescope Institute
10:30	**Break**	

11:00	**Session 1: A New Vision for K-12 Science and Engineering Education and NASA SMD Education**	*Moderator:* Brett Moulding, Utah Partnership for Effective Science Teaching and Learning

Guiding Questions and Focus:

- Present an overview of the NGSS and the role of NASA in supporting science and engineering education.
- How can/does NASA interact effectively with the education system to support K-12 science and engineering education?
- What opportunities does NASA SMD have to better support the new vision described in the NRC's *A Framework for K-12 Science Education*?
- How can/does NASA integrate the science and engineering talent of NASA SMD into the SMD education programs?

	Keynote Presentation	Stephen Pruitt, Achieve
	Panel Discussion	*Panelists:* Maya Garcia, Office of the State Superintendent of Education, D.C. John Ristvey, University Corporation for Atmospheric Research Holly Ryer, Space Telescope Science Institute Sam Shaw, South Dakota Department of Education
	Audience Joins the Discussion	
12:15 p.m.	**Lunch**	
1:30	**Session 2: Space Science Education Curriculum and Materials**	*Moderator:* Richard McCray, NAS, University of California, Berkeley

Guiding Questions and Focus:

- How do the instructional strategies advocated for in the NASA education programs match the Vision for Science Education described in the NRC's *Framework for K-12 Science Education*?
- How can NASA best encourage and support teachers to use NASA education resources in the classroom?
- What is the mechanism by which NASA education programs' instructional content material will be aligned to the Framework and the NGSS?
- How will NASA programs measure how well NASA EPO materials align to the NGSS?
- Information technology is changing the way science is done (data mining and simulations, for example)—What new possibilities does this development raise for the science classroom?

	Keynote Presentation: *Bringing Space Down to Earth and into the Classroom*	Edna DeVore, SETI Institute
	Panel Discussion	*Panelists:* Beth Johnston, Principal at Endeavour Elementary School Mordecai Mac Low, American Museum of Natural History Cassandra Soeffing, Institute for Global Environmental Strategies Belinda Wilkes, Chandra X-Ray Center
	Audience Joins the Discussion	

3:00 **Break**

3:30 **Session 3: Collaboration Among NASA SMD and K-12 Districts, Schools, and Teachers**

Moderator: Mitchell Nathan, University of Wisconsin, Madison

Guiding Questions and Focus:

- What are institutional arrangements that provide effective platforms for facilitating successful collaborations?
- How are evidence-based models for successful collaborations or partnerships being communicated across NASA education programs?
- How are proven models or strategies for scaling up and sustaining collaborations and partnerships being used in the NASA education programs?
- What are the barriers to accomplishing common goals across collaborating organizations? How can these barriers be overcome?

Keynote Presentation: ***Building the STEM Partnership Toolkit: Choosing Your Spots Carefully, Measuring Twice, and Finding Your Spanner When You Need It***

Gordon Kingsley, Georgia Tech

Poster Session

Presenters:
Nancy Ali, Space Sciences Laboratory
Lindsay Bartolone, Southwest Research Institute
Lin Hartung Chambers, NASA Langley Research Center
Troy Cline, NASA Goddard Space Flight Center
Anita Davis, Sigma Space
Bonnie Eisenhamer, Space Telescope Science Institute
Dorian Janney, NASA Goddard Space Flight Center
Andrea Jones, NASA Goddard Space Flight Center
Sheri Klug-Boonstra, Arizona State University
Keliann LaConte, Lunar and Planetary Science Institute
Kathleen Lestition, Smithsonian Astrophysical Observatory, Chandra X-Ray Center
Nancy Maryboy and David Begay, Indigenous Education Institute
Tony Murphy, The GLOBE Program
Luisa Rebull, NASA/IPAC Teacher Archive and Research Program
Daniella Scalice, NASA Astrobiology Program

Panel Discussion

Panelists:
Kathryn Flanagan, Space Telescope Science Institute
James Lochner, Universities Space Research Association
Lora Bleacher, NASA GSFC

Audience Joins the Discussion

5:30 **Workshop Adjourns for the Day**

5:30 **Reception**

DECEMBER 3, 2014

9:00 a.m. **Welcome** Committee Co-Chairs
Summary of Day 1

9:15 **Session 4: Supporting Science and Engineering Teachers through Professional Development** *Moderator:* Albert Byers, National Science Teachers Association

Guiding Questions and Focus:
- How are standards for professional development used in NASA professional development programs?
- How do the mechanisms and programs by which NASA programs meet the needs of in-service teachers, and how does this differ from the ways NASA programs meet the needs of pre-service teachers?
- What are the most effective and widely used delivery models (online, train the trainers, professional learning communities, summer seminars, internships) for NASA professional development programs?
- What are example strategies for partnering scientists and educators?

Keynote Presentation: ***Preparing Teachers to Support Three-Dimensional Science Learning*** Bill Penuel, University of Colorado, Boulder

Panel Discussion *Panelists:*
Annette DeCharon, University of Maine
Sheri Klug-Boonstra, Arizona State University
Mariel Milano, Orange County Public Schools, Florida

Audience Joins the Discussion

10:45 **Break**

11:15 **Session 5: Part 1—Evaluation of Education** *Moderator:* Theresa Schwerin, Institute for Global Environmental Strategies

Guiding Questions and Focus:
- What are current leading theories of STEM education evaluation (e.g., evidence-based, logic models)?
- Selecting the most appropriate assessment(s) for a given situation is a common challenge across education evaluation. What are leading factors or best practices that you recommend in selecting the most appropriate assessment(s) for a given situation?
- The goal of many STEM-related professional development efforts is to facilitate a change or increase in teacher effectiveness that in turn increases student learning. Additionally many efforts aspire to increase students' awareness/interest in STEM careers or students' desire to pursue more STEM-related coursework during high school and college. These impacts are sometimes challenging to capture as part of an evaluation. What are some methods or data that might shed light on these often elusive goals?
- What can we realistically measure? What can't we?
- What are the attributes of the evaluation tools that are consistent with effective evaluation of education programs?

	Keynote Presentation: ***Making the Right Choices: How to Get the Most Value out of eVALUation!***	Martin Orland, WestEd, and Steve Schneider, WestEd
12:00 p.m.	**Lunch**	
1:15	**Session 5: Part 2—Evaluation in Practice within NASA SMD**	*Moderator:* Theresa Schwerin, Institute for Global Environmental Strategies

Guiding Questions and Focus:

- Why and how does NASA evaluate the programs it executes?
- What are examples of evidence that the evaluation of NASA's programs is providing useful information to improve the programs?
- How does NASA make a difference in STEM education, and how is this known?
- What are the greatest challenges or barriers that people have encountered related to SMD education evaluation? What strategies have been used or recommended for addressing these barriers?
- How does the evaluation of NASA programs compare to the model presented for education by the speaker in Part 1 of this session?
- What is the mechanism by which the results of evaluation change NASA education programs?

	Keynote Presentation: ***Using Evaluation to Increase and Measure the Impact of Education***	Hilarie Davis, TLC Inc.
	Panel Discussion	*Panelists:* Bonnie Eisenhamer, Space Telescope Science Institute Jenny Gutbezahl, Brandeis University Frances Lawrenz, University of Minnesota
	Audience Joins the Discussion	
2:30	**Session 6: Enabling Actions** • Engage the audience in breakout groups related to each of the previous sessions • Instructions • Discussion/Breakout Groups Meet	*Moderator:* James Manning, Education Consultant

Breakout 1: Aligning to Standards

- What actions can NASA take to build upon, leverage, and/or expand its current efforts to align to and support the new vision described in the NRC's *A Framework for K-12 Science Education*, NGSS, and other standards initiatives?
- What new opportunities can be explored, and what challenges need to be overcome?

Breakout 2: Curriculum Support Resources

- What actions can NASA take to build upon, leverage, and/or expand its current efforts to translate its science into curriculum support materials and resources for formal and informal education and encourage educator use?
- What new opportunities can be explored, and what challenges need to be overcome?

Breakout 3: Collaborations

- What actions can NASA take to build upon, leverage, and/or expand its current collaborations among scientists, teachers, and formal and informal education institutions?
- What new opportunities can be explored, and what challenges need to be overcome?

Breakout 4: Professional Development

- What actions can NASA take to build upon, leverage, and/or expand its current efforts to provide professional development support to pre-service and in-service teachers and informal educators?
- What new opportunities can be explored, and what challenges need to be overcome?

Breakout 5: Evaluation

- What actions can NASA take to build upon, leverage, and/or expand its current efforts in measuring and assessing its impact in science and engineering education?
- What new opportunities can be explored, and what challenges need to be overcome?

3:15	**Break**	
3:30	**Session 6: Enabling Actions, *continued***	*Moderator:* James Manning, Education Consultant
	Reporting of Group Discussions	
4:45	**Summary and Wrap Up**	Committee Co-Chairs
5:00	**Workshop Adjourns**	

C

Poster Abstracts

THEMIS GEONS Magnetometer Program: Sustaining Teacher Engagement in NASA Science for over a Decade

Nancy Alima Ali, Coordinator of Education Programs, Multiverse, Space Sciences Laboratory, University of California, Berkeley

The Geomagnetic Event Observation Network by Students (GEONS) program was initiated by the THEMIS mission Education/Public Outreach team in 2004. The GEONS program trained science teachers across the United States on how to teach heliophysics using data collected from mission-related magnetometers that were installed in their schools. Teachers who participated in the GEONS program went on to become engaged in other NASA Science Mission Directorate professional development programs such as the Heliophysics Educator Ambassador program and the Heliophysics Community of Practice. This poster discusses the evidence behind the success of the GEONS program in establishing long-term partnerships between NASA scientists, teachers, research labs and informal education organizations. It addresses evaluation-based strategies for overcoming barriers between collaboration across organizations.

Space Explorers Club and Heliophysics Educator Ambassadors: Growing District and Teacher Partner Relationships for Sustainable and Significant Impacts

Lindsay Bartolone, Education and Public Outreach Lead, NASA's Interstellar Boundary Explorer (IBEX) Mission

Sustained funding through NASA's Science Mission Directorate for nearly a decade from the Interstellar Boundary Explorer Mission (IBEX) allowed rich partnerships between formal, informal and scientific institutions to develop, strengthen, become more efficient and grow effectively. Initial funding allowed for the collaborative development of the GEMS Space Science Sequence for grades 6-8, a product of partnership between NASA scientists and E/PO professionals with Lawrence Hall of Science, Adler Planetarium and NASA Science Mission Directorate Forums. These curricular units were nationally field-tested with local teacher partners and the curricular model was also tested against a control instructional method and showed greater positive cognitive and affective gains in students. These materials, complemented by additional NASA SMD educational support materials, were used in a variety of collaborative educator professional development programs including the Heliophysics Educator Ambassador program (train-the-trainer and Community of Practice strategies), and the Space Explorers Afterschool Club program, which additionally utilized partnerships with local school distircts, other NASA missions, and partner institutions and teachers. This poster will describe the partnerships, programs, connections and evidence for success of these developing models.

Mars Education: Providing an Evidence-Based Model for Authentic, STEM-Practice-based Learning

Dr. Catherine Bowman, Raytheon
Michelle Viotti, Jet Propulsion Laboratory
Sheri Klug-Boonstra, Arizona State University

NASA's Mars Exploration Program (MPE) has an evidence-based model for engaging teachers and students in authentic, standards-aligned STEM practices, using NASA data and discipline-based tools. Per recommendations from a large-scale external evaluation (SAMPI, 2011), all are guided by NSF's "Framework for Evaluating Impacts of Informal Science Education Projects" (Friedman, 2008), with logic models and impacts/indicators tables. Reviewed and refined annually, they guide ongoing internal evaluation. Project instruments (questionnaires, interview protocols, rubrics) are developed and tested following evidence-based procedures (e.g. Dillman, Smyth, & Christian, 2009; Lantz, 2004; Maxwell, 2005; Strauss & Corbin, 1998). The latest multi-year evaluation (Bowman, 2014) indicates MPE meets the majority of logic-model-based outcomes, and provides recommendations for the next evaluation phase, contributing to the nation's research base on effective STEM education models.

Sharing the Adventure: Observation-Based and Data-Based Examples

Lin Hartung Chambers, Director, CERES S'COOL Project, NASA's Langley Research Center

The NASA Langley Research Center Science Directorate houses two long-running projects aimed at involving students with the excitement of NASA's Earth science research. S'COOL is an observation project with a strong interactive aspect, while MY NASA DATA has a data focus. Both projects are integrally tied to NASA science and missions and involve collaboration across organizations at multiple levels: the project teams themselves involve scientist-educator-technologist collaborations; the projects engage teachers and students with NASA; and the projects also involve collaborations across NASA missions and Centers; and even to outside organizations and other agencies. Both projects have also encountered some barriers to optimal collaboration. This poster will introduce the two projects and explore the benefits and challenges of collaboration.

Building Digital Age Resources Through Sustained Partnerships: MMS and ISTE

Troy Cline, Education and Public Outreach Lead, Magnetospheric Multiscale Mission, NASA's Goddard Space Flight Center

This poster session will show how the Magnetospheric Multiscale (MMS) Mission contributes to teacher professional development opportunities in partnership with the International Society for Technology in Education (ISTE). ISTE is the trusted source in educational technology for professional development, knowledge generation, advocacy, and leadership for innovation and represents more than 100,000 professionals worldwide. The MMS team works closely with ISTE to design, develop and disseminate STEM educational materials to K-12 teachers throughout the life of the mission. Featured projects include: The Cyber Café (an online collaborative teacher workspace), a Computational Thinking Student Activity e-book and companion teacher guide iBooks embedded with self-paced professional development tools, information, and resources. ISTE has a proven record of accomplishment and sustainability and is well positioned to help transform education to meet the needs of students in the Digital Age.

Best Practices from the Earth to Sky Interagency Partnership

Anita Davis, Lead, Earth to Sky Interagency Partnership, SSAI at NASA Goddard Space Flight Center
Ruth Paglierani, Multiverse, University of California, Berkeley
Sandy Spakoff, National Conservation Training Center, U.S. Fish and Wildlife Service
John Morris, Alaska Region, National Park Service

The Earth to Sky interagency partnership is a ten-year long effort that has grown from its inception as a week-long NASA workshop for informal educators into a national-scale partnership between NASA, National Park Service, US Fish and Wildlife Service and NOAA. The partnership has designed and executed a variety of professional development events for informal educators, in which NASA scientists have participated as presenters and as science advisors. In turn, scientists have grown in their ability to effectively communicate their work to our audience. Alumni from the partnership together with science presenters form a growing community of practice of about 500 individuals. Independent evaluation determined that 86 alumni from two courses in turn reached over 4 million visitors to parks and refuges with content derived from ETS courses, through a variety of products and programs ranging from news articles to exhibits, peer and teacher-training events, ranger-led interpretive programs, and much more.

We will illustrate the best practices of ETS that have enabled us to build, maintain, deepen and expand a successful and effective collaboration between NASA scientists, and the educational institution that is represented by the training divisions and the professional informal educators of both NPS and USFWS. The complementary skills, resources, facilities and programs of the contributing agencies that have made possible a fruitful and sustainable relationship will be illustrated, and the characteristics of each agency partner's role in the leadership of ETS will be outlined. We will describe the means by which the ETS model has been shared within NASA and its associated educational communities, and the degree to which the ETS project is being accessed and used by NASA education and outreach. Suggestions for addressing challenges to achieving success in cross and within-agency collaboration will be included.

Strategic Partnerships: The Key to Sustainability and Reach for SMD Education

Bonnie Eisenhamer, Office of Public Outreach, Space Telescope Science Institute
The Space Telescope Science Institute (STScI) Education Team

STScI is the home institution for the education and public outreach activities of the Hubble and future James Webb space telescopes. Over time, STScI's Office of Public Outreach has established the infrastructure needed for reaching various audiences at the local, regional, and national levels. Partnerships are a critical element of this infrastructure, and sustainability of our program is ensured through our ongoing partnerships with organizations and institutions with staying power and reach. We have learned from past efforts that strategic partnerships can foster innovation, support diversity initiatives, and increase impact in a cost-effective way while providing target audiences with greater access to NASA SMD science and resources. Partners are selected based upon specific criteria such as potential for reach, the percentage of underrepresented educators and students served, complementary program goals, and willingness to collect and share evaluation data and results with us. This poster will highlight our partnership model as well as examples and benefits of strategic partnerships over time.

Empowering Educators to Engage with NASA Mission Science

Dorian Janney, Formal Education Specialist, Global Precipitation Measurement Mission, NASA Goddard Space Flight Center

In February, 2014, NASA launched the Global Precipitation Measurement (GPM) mission. This satellite is now measuring precipitation as it falls all over the globe, and is able to update the data every three hours. This poster will share information about some of the education outreach projects that were developed to share the science and technology behind this mission. The "GPM Master Teacher Program" will be the focus as it is a highly successful model for engaging and collaborating with educators around the world using online tools. GPM's science team members serve as experts as they deliver background science and engineering content to the educators during the monthly webinars. Metrics are collected to determine the effectiveness of using NASA-unique educational resources to bring STEM content and teach NGSS to students in formal education settings

LRO's Lunar Workshops for Educators: A Proven Model for Exceptional Teacher Professional Development

Andrea Jones, Education Specialist, Planetary Science Institute, NASA Goddard Space Flight Center
Lora Bleacher, Education and Public Outreach Lead, Solar System Exploration Division, NASA Goddard Space Flight Center
Sanlyn Buxner, Education Specialist, Planetary Science Institute
Marti Canipe, Graduate Student in Science Education, University of Arizona

The Lunar Workshops for Educators is an award-winning series of weeklong professional development workshops for grade 6–9 science teachers, sponsored by the Lunar Reconnaissance Orbiter (LRO) and conducted by the LRO Education Team. We will present an overview of the workshop series, a summary of five years of evaluation results, and highlight the strengths of this model for educator professional development, in a program made possible through content and assets unique to NASA's Science Mission Directorate.

Explore! Engaging Children in Space Science in Libraries and Other Out-of-school Programming

Keliann LaConte, Informal Education/Explore! Program Lead, Lunar and Planetary Science Institute

The Lunar and Planetary Institute's *Explore* program builds the capacity of informal educators—especially librarians—to provide Earth and space science and engineering experiences for children and families utilizing unique NASA Science Mission Directorate assets. Hands-on activities, programming resources, and training are developed in collaboration between scientists, evaluators, librarians, and professional library organizations through an advisory board, formative and summative evaluation and research, and field tests. Evaluation and research data show that training participants gained statistically significantly in Earth and space science content knowledge, experienced an increase in confidence and self-reported ability and intention to use the activities, and are actively using *Explore* materials. Ongoing communication via *Explore* and partner networks and frequent follow-up opportunities are crucial for sustaining librarian and professional library organization involvement.

NASA's Chandra X-ray Observatory Education, Public Engagement and Communications Program

Kathy Lestition, Education/Outreach Coordinator, Chandra X-Ray Center, Smithsonian Astrophysical Observatory

NASA's Chandra X-ray Observatory Education, Public Engagement & Communications program utilizes established working relationships with scientists to provide the starting point for all Chandra educational materials. Partnering with organizations such as the National Science Olympiad, the 4-H, the NASA Museum Alliance and the American Library Association, among others, leverages external distribution networks for national impact. Enabling and sustaining a network of "volunpeers" empowered to organize science education events in their communities further strengthens the reach of educational science materials. We summarize a sample of our synthesized suite of programs in informal and formal education that communicate the compelling topics that the high-energy Universe can reveal as well as provide an overview of the guiding research and evaluation results.

Indigenous Education Institute: Collaboration with Integrity

Nancy C. Maryboy, President and Executive Director, Indigenous Education Institute
David Begay, Adjunct Faculty, Department of Physics and Astronomy, Northern Arizona University

Indigenous Education Institute: Collaboration with Integrity will feature images and text focused on our more than 20 year history of working in cross-cultural settings, on collaborative projects funded by NASA and the National Science Foundation. We will show the importance of multicultural science education, offer examples of collaborative success, and strategies for scaling up and sustainability. The poster will showcase lessons learned, as our Indigenous institute has collaborated with science centers, NASA affiliates, universities, tribal colleges, museums, schools and planetariums. We will highlight current NASA projects such as Navajo Sky—Education Modules for Digital Planetariums, and MAVEN—Imagine Mars Through Indigenous Eyes.

Connecting GLOBE Students to Satellite Mission Partnerships

Tony Murphy, Director, Global Learning and Observations to Benefit the Environment (GLOBE) Program

Along with the ongoing CloudSat and CALIPSO missions, The GLOBE Program is involving educators and students in activities related to two new NASA satellite missions in 2014/15:

- Global Precipitation Measurement (GPM), launched in February 2014; and
- Soil Moisture Active Passive (SMAP), scheduled for launch in January 2015.

Instruments on the various satellites take measurements of clouds, aerosols, atmospheric chemistry, precipitation, soil moisture and other elements critical to understanding Earth's changing climate.

Satellite partnerships provide students and teachers the opportunity to contribute to the science of the mission by collecting environmental data on Earth to compare to that of the Earth-orbiting satellites. The partnerships also offer a range of opportunities for collaboration: students with other students; students with teachers; students and teachers with scientists; and cooperating agencies with one another in support of the missions.

NASA/IPAC Teacher Archive Research Program (NITARP): Evidence of a Successful Partnership Over a Decade

Luisa Rebull, Director and Mentor Scientist, NASA/IPAC Teacher Archive and Research Program
G. K. Squires, Assistant Astronomer, Institute for Astronomy, University of Hawaii
V. Gorjian, Research Astronomer, JPL; The NITARP Team

NITARP provides educators with an authentic astronomical research experience by partnering small groups of educators from across the U.S. with a mentor astronomer for a year-long original research project, during which the teams echo the entire research process: writing a proposal, conducting research, writing up and presenting the results at an American Astronomical Society (AAS) conference. Few science teachers have had an authentic research experience. Our goal is to expose teachers to this process, as messy as it can be, but also rewarding, to help them understand what a career in research would entail, and better prepare their students. The most recent formal evaluation looked at both the cognitive and affective impacts of NITARP on teachers, and extent to which NITARP changes teaching styles, or creates a desire to teach science differently.

NASA Astrobiology Institute: Embedding E/PO in the Place of Science to Maximize Collaboration, Partnerships, and Impact

Daniella Scalice, Education and Public Outreach Lead, NASA Astrobiology Program

The NASA Astrobiology Institute (NAI) is a nationally distributed, interdisciplinary institute-without-walls that brings together teams of scientists to study life's origins and the possibility of life elsewhere in the Universe. NAI's broad research portfolio is conveyed to learners of all ages in a variety of programs by E/PO leads embedded in each research team. This model maximizes the participation of the astrobiologists themselves—key to achieving participant outcomes—and also allows for long-term partnerships with local institutions such as schools, libraries, and museums—key to designing programs responsive to learners' needs.

NAI E/PO leads are coordinated by a central office, which facilitates communication and collaboration, and administers services such as supplemental funds and evaluation support. As grantees, the NAI E/PO leads are able to deliver their competitively-selected activities with a supportive, centralized presence; as a community of practice, the NAI E/PO leads are able to collaborate, share resources, leverage partners, and replicate successful strategies.

D

Workshop Participants

Nancy Ali, Space Sciences Laboratory
Marc Allen, NASA
Laura Angle, Students4SSTEM
Lindsay Bartolone, Southwest Research Institute
David Begay, Indigenous Education Institute
Lora Bleacher, NASA Goddard Space Flight Center (GSFC)
Eric Brown de Colstoun, NASA
Albert Byers, National Science Teachers Association
Jordan Camp, NASA GSFC
Valerie Casasanto, University of Maryland Baltimore County (UMBC), Joint Center for Earth Systems Technology/NASA GSFC
Lin Chambers, NASA Langley Research Center (LaRC)
Philip Christensen, Arizona State University
Troy Cline, NASA GSFC (Education and Public Outreach Lead for the Magnetospheric Multiscale Mission)
Emily CoBabe-Ammann, University Corporation for Atmospheric Research (UCAR), Community Programs
Sarah Crecelius, Science System and Applications, Inc./NASA LaRC
Maizie Cummings-Rocke, Rockville Science Center
Hilarie Davis, TLC, Inc.
Annette DeCharon, University of Maine
Belay Demoz, UMBC
Edna DeVore, SETI Institute
Charles Divine, Divine Solutions
Bonnie Eisenhamer, Space Telescope Science Institute (STScI)
Kristen Erickson, NASA Science Mission Directorate (SMD)
Sarah Eyermann, Syneren Technologies; NASA GSFC
Carmen Fies, University of Texas, San Antonio
Kathryn Flanagan, STScI
Gloria Fulwood, Smithsonian National Air and Space Museum (NASM)
Robert Gabrys, NASA GSFC

Maya Garcia, Office of the State Superintendent of Education (OSSE), Government of the District of Columbia
Joel Green, STScI
Katia Grigoriants, OSSE, Government of the District of Columbia
Jenny Gutbezahl, Brandeis University
Carol Haden, Magnolia Consulting
Heidi Hammel, Association of Universities for Research in Astronomy
Wesley Harris, Massachusetts Institute of Technology
Charles Harris, U.S. House of Representatives, Committee on Science
Hashima Hasan, NASA Headquarters
Mary Haskins, Rockhurst University
Grace Hu, Office of Management and Budget
Dorian Janney, NASA GSFC
Hussein Jirdeh, STScI
Beth Johnston, Endeavour Elementary School
Michael Johnston, spouse of workshop participant
Andrea Jones, Planetary Science Institute
Dean Kern, NASA GSFC
Gordon Kingsley, Georgia Institute of Technology
Sheri Klug-Boonstra, Arizona State University
Keliann LaConte, Lunar and Planetary Institute (LPI)
Frances Lawrenz, University of Minnesota
Kathy Lestition, Smithsonian Astrophysical Observatory, Chandra X-ray Center
Ellen Lettvin, Department of Education
Elaine Lewis, NASA
Preston Lewis, NASA LaRC
Miles Lifson, Aerospace Industries Association
James Lochner, Universities Space Research Association
Ota Lutz, Jet Propulsion Laboratory
Mordecai Mac Low, American Museum of Natural History/Columbia University
Stephen Mackwell, LPI
James Manning, Education Consultant
John Maple, STScI
Nancy Maryboy, Indigenous Education Institute
John Mather, NASA GSFC
Barbara Mattison, Universities Space Research Association (USRA)/Center for Research and Exploration in Space Science and Technology
Louis Mayo, NASA
Richard McCray, University of California, Berkeley
Jennifer McIntosh, Smithsonian Institution
Kevin McLin, Sonoma State University, Education and Public Outreach Group
Bonnie Meinke, STScI
Mariel Milano, Orange County Public Schools, Florida
Sarah Mitchell, Syneren Technologies/NASA GSFC
Brett Moulding, Utah Partnership for Effective Science Teaching and Learning
Tony Murphy, GLOBE Program
Mitchell Nathan, University of Wisconsin, Madison
Dava Newman, Massachusetts Institute of Technology
Daniel Newmyer, Triangle Coalition/NASA GSFC, Office of Education
Sten Odenwald, National Institute of Aerospace
Daniel Oostra, NASA

Martin Orland, WestEd
Bill Penuel, University of Colorado, Boulder
Laura Peticolas, University of California, Berkeley
Kara Pezzi, Albert Einstein Distinguished Educator Fellow at the Department of Energy
Miloslawa Piszczek, FIRST
Stephen Pruitt, Achieve
Juliet Ray, University of Texas, San Antonio
Luisa Rebull, Spitzer Science Center
Ann Reimers, Department of Energy
Holli Riebeek, NASA GSFC
John Ristvey, UCAR
Holly Ryer, STScI
Daniella Scalice, NASA Ames Research Center
Deborah Scherrer, Stanford University
Greg Schultz, Astronomical Society of the Pacific
Theresa Schwerin, Institute for Global Environmental Strategies
Adrianne Sethi, Smithsonian NASM
Sam Shaw, South Dakota Department of Education
Linda Sherman, NASA Wallops Flight Facility Education
Josh Shiode, American Astronomical Society
Stephanie Shipp, LPI
Linda Shore, Astronomical Society of the Pacific
Denise Smith, STScI
Cassie Soeffing, Institute for Global Environmental Strategies
Tim Spuck, Associated Universities, Inc.
Beverly Stambaugh, Albert Einstein Distinguished Educator Fellow
David Stern, Indigenous Education Institute
Stephanie Stockman, NASA Headquarters
Mary Styers, Magnolia Consulting
Frank Summers, STScI
Terry Teays, Johns Hopkins University/Maryland Space Grant Consortium
Ronni Teich, Macalester College
Michelle Thaller, NASA GSFC
Mark Thiemens, University of California, San Diego
Dawn Turney, Johns Hopkins University/Applied Physics Laboratory
Elizabeth Wallace, Starry Telling
Kristen Weaver, NASA GSFC; USRA; Goddard Earth Sciences Technology and Research
Ming-Ying Wei, NASA Headquarters
Alice Wessen, NASA Science, Planetary Division, Communications
Belinda Wilkes, Chandra X-Ray Center
Steven Williams, Smithsonian NASM
Joyce Winterton, NASA GSFC Wallops Flight Facility

E

Biographical Information for Organizing Committee and Staff

PHILIP R. CHRISTENSEN, *Co-Chair*, is a Regents Professor at Arizona State University. He is also the Ed and Helen Korrick Professor in the School of Earth and Space Exploration. His research interests focus on the composition, processes, and physical properties of Mars, Earth, and other planetary surfaces. Dr. Christensen uses spectroscopy, radiometry, field observations, and numerical modeling to study the geology and history of planets and moons. A major facet of his research is the design and development of spacecraft instruments. He has built five science instruments that have flown on NASA's Mars Observer, Mars Global Surveyor, Mars Odyssey, and the Mars Exploration Rover missions. Over the past decade he has studied urban environments and growth worldwide using satellite data and has developed an extensive K-12 education and outreach program to bring the excitement of science and exploration into the classroom. Dr. Christensen was awarded NASA's Exceptional Scientific Achievement Medal in 2003, NASA's Public Service Medal in 2005, and the G.K. Gilbert Award of the Geological Society of America in 2008. He is a fellow of the American Geophysical Union (AGU) and the Geological Society of America. He received his Ph.D. in geophysics and space physics from the University of California, Los Angeles. Dr. Christensen currently serves as co-chair of the National Research Council's (NRC's) Committee on Astrobiology and Planetary Science and has previously served as a member of the NRC Committee on Planetary and Lunar Exploration and as chair of the Planetary Science Decadal Survey's Mars Panel.

BRETT D. MOULDING, *Co-Chair*, is the director of the Utah Partnership for Effective Science Teaching and Learning. Mr. Moulding was the state science education specialist and coordinator of curriculum and then director of the curriculum and instruction before retiring in 2008. Mr. Moulding taught chemistry for 20 years at Roy High School in the Weber School District and served as the district teacher leader for 8 years. Mr. Moulding also served on the Board at the Triangle Coalition, the NAEP 2009 Framework Committee, and was the President of the Council of State Science Supervisors from 2003 to 2006. He was a member of the NRC's Board on Science Education and a member of the committee that authored *A Framework for K-12 Science Education*. He subsequently served as a writing team leader for Achieve's Next Generation Science Standards. Mr. Moulding has received the Governor's Teacher Recognition Award, the Presidential Award for Excellence in Mathematics and Science Teaching, and the Award of Excellence from the Governor's Science and Technology Commission. He graduated from the University of Utah with a B.S. in chemistry with minors in biology, math, and physics. He has a M.S. in curriculum and instruction from Weber State University and an Administrative Supervisory Certificate from

Utah State University. Most recently, he has served on the NRC's Committee on Conceptual Framework for New Science Education Standards and the Board on Science Education.

ALBERT BYERS is the associate executive director of services at the National Science Teachers Association (NSTA). Prior to NSTA, Dr. Byers worked as an aerospace education specialist at NASA Goddard Space Flight Center (GSFC). At NSTA, Dr. Byers provides strategic oversight and executive management of NSTA services that include the production, delivery, and evaluation of large-scale blended learning solutions that include NSTA conferences, competitions, summer academies and institutes, the NSTA Learning Center e-learning portal, SciLinks, and government partnerships. The NSTA Learning Center has more than 150,0000 active users spending hours online each week in self-directed learning experiences in collaboration with 90 school districts and 39 universities across the country. NSTA's five annual conferences reach 18,000-20,000 individuals per year across 16 programming days, and teacher and student competitions reach tens of thousands annually. Dr. Byers serves on the primary technical working group for the Department of Education's online communities of practice efforts, and was previously a delegate for the Department of Education at the Asian Pacific Economic Cooperation Summit, and as an expert panelist for the National Assessment of Education Progress Science Framework Prioritization working group. Dr. Byers earned his Ph.D. in science education from Virginia Polytechnic Institute and State University.

HEIDI B. HAMMEL is executive vice president at the Association of Universities for Research in Astronomy (AURA). Previously she served as a senior research scientist with the Space Science Institute, an independent research and education organization based in Boulder, Colorado. Her primary research interests are the outer planets and their satellites, with a specific focus on observational techniques. Dr. Hammel is a leading expert on the planet Neptune and was a member of the Imaging Science Team during the Voyager 2 spacecraft's encounter with that planet in 1989. For the impact of Comet Shoemaker-Levy 9 with Jupiter in July 1994, Dr. Hammel led the Hubble Space Telescope (HST) team that investigated Jupiter's atmospheric response to the collisions. Her latest research has focused on the imaging of Neptune and Uranus with the HST and on ground-based observations of Uranus. Dr. Hammel is the recipient of many awards, including the 1996 Urey Prize from the American Astronomical Society's Division for Planetary Sciences (AAS/DPS) and the San Francisco Exploratorium's 1998 Public Understanding of Science Award. More recently, Dr. Hammel was elected a fellow of the American Association for the Advancement of Science in 2000 and received the AAS/DPS's Sagan Medal for outstanding communication by an active planetary scientist to the general public in 2002. In addition, Asteroid 1981 EC20 has been renamed 3530 Hammel in her honor. She received her Ph.D. in astronomy and astrophysics from the University of Hawaii. She has served on the NRC Committee on Planetary and Lunar Exploration and on the Panel on Solar System Exploration of the Committee on Priorities for Space Science Enabled by Nuclear Power and Propulsion: A Vision for Beyond 2015. In addition to serving on the Space Studies Board (SSB), Dr. Hammel served as chair of the Giant Planet Panel of the NRC's Planetary Science Decadal Survey and most recently as a member of the SSB.

WESLEY L. HARRIS is the Charles Stark Draper Professor of Aeronautics and Astronautics and associate provost at Massachusetts Institute of Technology. He is a former NASA associate administrator for aeronautics responsible for all aeronautics programs, facilities, and personnel (1993-1995). Prior to NASA, he was the University of Tennessee Space Institute's vice president and chief administrative officer. His academic research with unsteady aerodynamics, aeroacoustics, rarefied gas dynamics, sustainment of capital asset, and chaos in sickle cell disease have made seminal contributions in these fields. As an elected fellow of the American Institute of Aeronautics and Astronautics and of the American Helicopter Society, Dr. Harris was recognized for personal engineering achievements, engineering education, management, and advancing cultural diversity. He is also an elected member of the National Academy of Engineering, the Cosmos Club, and the Confrerie des Chavaliers du Tastevin. He earned his Ph.D. in aerospace and mechanical sciences from Princeton University. Dr. Harris currently serves on the NRC's Laboratory Assessments Board and the Panel on Mechanical Science and Engineering at the Army Research Laboratory, and was the chair of the Committee on Review of Army Research Laboratory Programs for Historically Black Colleges and Universities and Minority Institutions.

CHARLES F. KENNEL is a distinguished professor of atmospheric science emeritus in the Scripps Institution of Oceanography at the University of California, San Diego (UCSD). Dr. Kennel was the founding director of the UCSD Environment and Sustainability Initiative, an all-campus effort embracing teaching, research, campus operations, and public outreach, and is now chair of its international advisory board. His research covers plasma physics, space plasma physics, solar-terrestrial physics, plasma astrophysics, and environmental science and policy. He is a member of the National Academy of Sciences, the American Academy of Arts and Sciences, the American Philosophical Society, and the International Academy of Astronautics. He was a member of the NASA Advisory Council from 1998 to 2006, its chair from 2001 to 2005, and is presently chair of the California Council on Science and Technology. He has had visiting appointments to the International Centre for Theoretical Physics (Trieste), the National Center for Atmospheric Research (Boulder), the Ecole Polytechnique (Paris), California Institute of Technology (Pasadena), Space Research Institute (Moscow), and the University of Cambridge (U.K.). He is a recipient of the James Clerk Maxwell Prize (American Physical Society), the Hannes Alfven Prize (European Geophysical Society), the Aurelio Peccei Prize (Accademia Lincei), and the NASA Distinguished Service and Distinguished Public Service Medals. He was the 2007 C.P. Snow Lecturer at Christ's College, Cambridge (U.K.). He earned his Ph.D. in astrophysical sciences from Princeton University. Dr. Kennel has served on numerous NRC committees and boards including the 2010 astronomy and astrophysics decadal survey follow-on—the Panel on Implementing Recommendations from New Worlds, New Horizons Decadal Survey. Dr. Kennel most recently served as chair of the NRC's SSB.

JAMES MANNING is a consultant. He was the former head of the Office of Public Outreach at the Space Telescope Science Institute and the former executive director of the Astronomical Society of the Pacific. Mr. Manning has spent 39 years working in science education and public outreach (EPO), in production, presentation, teaching, and administration. For nearly three decades he worked at planetariums at the University of North Carolina, Parkland College, and Montana State University/Museum of the Rockies, developing and opening the latter two and serving as their director. His work included planetarium production and presentation, exhibit and program development, teacher professional development, and facility management. He also served a term as president of the International Planetarium Society. He has served on the astrophysics subcommittee of the NASA Advisory Council, and he continues to consult on NASA Science Mission Directorate (SMD) Education and Public Outreach efforts through the Astrophysics Science Education and Public Outreach Forum. He earned his M.A. in physics and astronomy from the University of North Carolina.

RICHARD A. McCRAY is a visiting scholar at the University of California, Berkeley. He is also the George Gamow Distinguished Professor of Astrophysics, emeritus, at the University of Colorado, Boulder (UCB). Dr. McCray's research is in the theory and observations of the dynamics of the interstellar gas and cosmic X-ray sources, supernovae, and supernovae remnants. Prior to UCB, he was an assistant professor at the Harvard College Observatory, and he has also held visiting positions at the NASA GSFC, Beijing University and Nanjing University, the Space Telescope Science Institute, and Columbia University. In 1983 Dr. McCray was awarded a Guggenheim Foundation Fellowship and in 1990 he received the Dannie S. Heinemann Prize for Astrophysics of the American Physical Society. In 1989 he was elected to the National Academy of Sciences, and his citation notes that "he is widely recognized as the world leader in theoretical x-ray astronomy." In 1996 he was appointed concurrent professor of astronomy at Nanjing University. In 2002 he was awarded the National Science Foundation (NSF) Director's Award for Distinguished Teaching Scholars. Dr. McCray earned his Ph.D. in theoretical physics from the University of California, Los Angeles. He is currently serving on the NRC's International Temporary Nominating Group for Class 1: Physical and Mathematical Sciences. His numerous prior committee service also includes the Board on Science Education, the Chair of Section: 12 Astronomy, and the SSB.

MITCHELL NATHAN is a professor in the Department of Educational Psychology at the University of Wisconsin, Madison (UW-Madison). At UW-Madison he also served as director of the Center on Education and Work, and as faculty member for the Latin American School for Education, Cognitive, and Neural Sciences. He holds affiliate appointments in the UW-Madison Department of Curriculum and Instruction, Department of Psychology, and

Wisconsin Center for Education Research. In research and development in artificial intelligence, computer vision, and robotic mobility, he has worked on the design and development of autonomous robotic arms and vehicles; the development of expert systems and knowledge engineering interview techniques; and the representation of perceptual and real-world knowledge to support inference making in dynamic environments. He has also worked on computer-based mathematics tutoring that relies heavily on students' comprehension processes for self-evaluation and self-directed learning (so-called unintelligent tutoring systems). Dr. Nathan directed the project Supporting the Transition from Arithmetic to Algebraic Reasoning (STAAR; funded by the Interagency Education Research Initiative, IERI), which studied the transition from arithmetic to algebraic reasoning. He is co-principal investigator for both the AWAKEN Project ("Aligning educational experiences with Ways of Knowing Engineering"), which documents how people learn and use engineering, and the National Center for Cognition and Mathematics Instruction. He is a member of the steering committee for the Delta Program, which promotes the development of a national faculty in the natural and social sciences, engineering, and mathematics that is committed to implementing and advancing effective teaching practices for diverse student audiences. He received his Ph.D. in experimental (cognitive) psychology and holds a B.S. in electrical and computer engineering, mathematics, and history. He previously served as a member of the NRC Committee Toward Integrated STEM Education: Developing a Research Agenda.

PATRICIA H. REIFF is professor in the Department of Space Physics and Astronomy at Rice University. Dr. Reiff has been involved in space plasma physics research for more than 40 years, with interests in magnetospheric convection, magnetosphere-ionosphere coupling, plasma particle acceleration mechanisms, and solar wind control of the magnetosphere and ionosphere of the Earth and Mercury. She was a co-investigator on the Dynamics Explorer, Polar, and IMAGE Missions, was a co-investigator for both science and public outreach for the PEACE electron spectrometer on Cluster II, is EPO lead for the MMS mission, and was on the EPO team for the Center for Integrated Space Weather Modeling. She provides real-time space weather alerts to about 1,000 subscribers. She has served as director for public education and teacher enhancement projects for more than 25 years. Her "Space Update" software has been used by more than 1 million visitors at more than 15 museums, and together with "Earth Update" and "Space Weather" has been distributed to more than 300,000 educators and learners. Her project "Immersive Earth," in cooperation with the Houston Museum of Natural Science, created a full-dome digital planetarium that shows teaching Earth science, and has created a portable planetarium system "Discovery Dome" to teach Earth and Space Science through immersive digital theater, that is now in more than 220 sites in 33 countries and 33 states. These projects have spun off two companies, Space Update, Inc. and MTPE, to distribute educational materials and portable planetariums. She has served as editor or associate editor for *EOS*, *Journal of Geophysical Research*, and *Reviews of Geophysics* and has served on the editorial board of *Space Weather*. She has served on advisory committees for the NSF, NASA, and the National Academy of Sciences, among others. She has served as chair of the Council of Institutions of the Universities Space Research Association and serves on the Space, Physics, and Aeronomy Public Education Committee for the AGU. She was elected to the Cosmos Club in 1992, was selected as a fellow of the AGU in 1997, and received the AGU "Athelstan Spilhaus Award" for public education in 2009. She received the "Aerospace Educator Award" from Women in Aerospace in 1999 and the Service Award from the Northwest Amateur Radio Society in 2004. She received NASA "Group Achievement" awards for the IMAGE, GGS, and Cluster missions, and was an organizer for the World Space Congress in 2002. In 2013 she received the first-ever "Space Physics and Aeronomy Richard Carrington" award from the AGU. She has led many teacher workshops and scientific tours. In addition to training 12 Ph.D.'s, she created a "Master of Science Teaching" degree, with 27 teacher alumni as of 2014. Dr. Reiff earned her Ph.D. in space physics and astronomy at Rice University. She has previously served as a member on the NRC Committee on Solar-Terrestrial Research and as chair of the Panel on the International Magnetospheric Study.

THERESA SCHWERIN is founding officer and vice president of education programs for the Institute for Global Environmental Strategies (IGES). She has more than 20 years of experience in the areas of science applications and education, communication, and information science. She leads IGES education initiatives, particularly the NASA Earth Science Education and Public Outreach Forum, IGES science contests for students, the Earth System Science Education Alliance, a NASA-NOAA-NSF sponsored project providing professional development for K-12 teachers, NASA Earth and Space Science Education Product Review, and a variety of NASA professional development and communication activities. She is also a key leader of nasawavelength.org—a digital library developed by IGES with the University of California, Berkeley. Wavelength provides robust tools for science educators—K-12, higher education, and out-of-school—for searching, browsing, and creating custom pathways and collections from NASA's SMD portfolio of more than 2,000 educational resources, in ways that are most meaningful for educators. Ms. Schwerin has led a wide range of new efforts for organizations such as NASA, NOAA, the United Nations, WGBH Education Foundation, and the former Japanese National Space Development Agency (now part of the Japanese Aerospace Exploration Agency, or JAXA). Her work has led to the planning, development, and implementation of international and national education programs, products, workshops, reviews, and conferences. Ms. Schwerin is a member of the American Geophysical Union, the NSTA, Astronomical Society of the Pacific, American Association for the Advancement of Science, and the American Library Association. Ms. Schwerin's related education experience also includes conducting children's programs in a public library system. Ms. Schwerin holds a M.S. in library and information science from the University of Maryland and a B.S. in sociology from the College of Charleston.

STAFF

ABIGAIL A. SHEFFER is a program officer for the SSB. In fall 2009, Dr. Sheffer served as a Christine Mirzayan Science and Technology Policy Graduate Fellow for the National Academies and then joined the SSB as an associate program officer. Since joining the NRC, she has been study director on reports such as *Landsat and Beyond—Sustaining and Enhancing the Nation's Land Imaging Program* and *The Effects of Solar Variability on Earth's Climate: A Workshop Report.* Dr. Sheffer has been an assisting staff officer on several other reports, including *Pathways to Exploration—Rationales and Approaches for a U.S. Program of Human Space Exploration* and *Solar and Space Physics: A Science for a Technological Society*. Dr. Sheffer earned her Ph.D. in planetary science from the University of Arizona and A.B. in geosciences from Princeton University.

MICHAEL MOLONEY is the director for Space and Aeronautics at the SSB and the Aeronautics and Space Engineering Board (ASEB) of the NRC of the National Academies. Since joining the ASEB/SSB, Dr. Moloney has overseen the production of more than 40 reports, including four decadal surveys—in astronomy and astrophysics, planetary science, life and microgravity science, and solar and space physics—a review of the goals and direction of the U.S. human exploration program, a prioritization of NASA space technology roadmaps, as well as reports on issues such as NASA's Strategic Direction, orbital debris, the future of NASA's astronaut corps, and NASA's flight research program. Before joining the SSB and ASEB in 2010, Dr. Moloney was associate director of the Board on Physics and Astronomy (BPA) and study director for the decadal survey for astronomy and astrophysics (Astro2010). Since joining the NRC in 2001, Dr. Moloney has served as a study director at the National Materials Advisory Board, the BPA, the Board on Manufacturing and Engineering Design, and the Center for Economic, Governance, and International Studies. Dr. Moloney has served as study director or senior staff for a series of reports on subject matters as varied as quantum physics, nanotechnology, cosmology, the operation of the nation's helium reserve, new anti-counterfeiting technologies for currency, corrosion science, and nuclear fusion. In addition to his professional experience at the National Academies, Dr. Moloney has more than 7 years' experience as a foreign-service officer for the Irish government—including serving at the Irish Embassy in Washington and the Irish Mission to the United Nations in New York. A physicist, Dr. Moloney did his Ph.D. work at Trinity College Dublin in Ireland. He received his undergraduate degree in experimental physics at University College Dublin, where he was awarded the Nevin Medal for Physics.

KATIE DAUD is a research associate for the SSB and the ASEB. Previously, she worked at the Smithsonian National Air and Space Museum's Center for Earth and Planetary Studies as a planetary scientist. Ms. Daud was a triple major at Bloomsburg University, receiving a B.S. in planetary science and Earth science and a B.A. in political science.

ANESIA WILKS is a senior program assistant. She brings experience working in the National Academies conference management office as well as other administrative positions in the D.C.-metropolitan area. Ms. Wilks has a B.A. in psychology (magna cum laude) from Trinity University.

MICHELLE THOMPSON is a Ph.D. student in planetary sciences at the University of Arizona's Lunar and Planetary Laboratory. Her research is focused on understanding the effects of space weathering on airless body surfaces. Ms. Thompson uses transmission electron microscopy to study microstructural and microchemical signatures of space weathering in lunar and asteroidal surface samples returned from the NASA Apollo missions and the JAXA Hayabusa mission. She has received several awards for her presentations at scientific conferences and was recently awarded a NASA Earth and Space Science Fellowship for her research. She serves on several committees as a student in Tucson, including as a representative for the graduate students to the faculty, coordinator for visiting colloquium speakers, and organizer of non-academic career seminars for the students in her department. She has been keenly interested in science policy since beginning graduate school and is very excited for the opportunity to work with the SSB. She looks forward to bringing her experiences at the SSB with her while pursuing a career in planetary science.

ANGELA DAPREMONT is an intern at the SSB. Ms. Dapremont developed an interest in the merging of science and policy as a result of participating in meetings with congressional aides about science education and funding during her final year of undergraduate study. Previously, she has conducted research in the fields of algal ecology, renewable energy, natural hazards, and planetary geology. She has participated in internships focused on Mars geology at NASA Johnson Space Center and NASA GSFC. She graduated from the College of Charleston with a B.S. in geology and a minor in French and francophone studies in May of 2014.